# Mutual Sustainability of Tubewell Farming and Aquifers

# Advances in Asian Human-Environmental Research



***Aims and Scope***

The series aims at fostering the discussion on the complex relationships between physical landscapes, natural resources, and their modification by human land use in various environments of Asia. It is widely acknowledged that human-environment-interactions become increasingly important in area studies and development research, taking into account regional differences as well as bio-physical, socio-economic and cultural particularities.

The book series seeks to explore theoretic and conceptual reflection on dynamic human-environment systems applying advanced methodology and innovative research perspectives. The main themes of the series cover urban and rural landscapes in Asia. Examples include topics such as land and forest degradation, glaciers in Asia, mountain environments, dams in Asia, medical geography, vulnerability and mitigation strategies, natural hazards and risk management concepts, environmental change, impacts studies and consequences for local communities. The relevant themes of the series are mainly focused on geographical research perspectives of area studies, however there is scope for interdisciplinary contributions.

For further volumes:
http://www.springer.com/series/8560

Ahmad Saeed Khattak

# Mutual Sustainability of Tubewell Farming and Aquifers

## Perspectives from Balochistan, Pakistan

Ahmad Saeed Khattak
Department of Geography
University of Balochistan
Quetta, Pakistan

ISSN 1879-7180 ISSN 1879-7199 (electronic)
ISBN 978-3-319-02803-3 ISBN 978-3-319-02804-0 (eBook)
DOI 10.1007/978-3-319-02804-0
Springer Cham Heidelberg New York Dordrecht London

Library of Congress Control Number: 2014942932



Printed on acid-free paper

Springer is part of Springer Science+Business Media (www.springer.com)

*Dedicated to*
*My Gracious Mother (Late)*
*&*
*My Wife – Gul Shameera*

# Preface

Agriculture serves the minimum and ultimate needs of mankind – food and shelter. In the current era of enhanced human needs for agro-products, and highly competing land-use economies, irrigated agriculture is considered the most promising option. Unfortunately, however, mankind often looks at fast-acting remedies to serve his needs and desires while ignoring their side effects. Likewise, in the exploitation of water resources for irrigation purposes, many societies have neglected sustainable lines of action. Consequently, many regions across the world are now faced with freshwater shortages, and are even at crisis level.

The Balochistan province of Pakistan is a typical subtropical dry region, where groundwater irrigated agriculture has been among the dominant income generators. As soon as modern deep drilling and pumping technologies disseminated here, the farmers rushed indiscriminately to exploit local aquifers. Consequently, tubewell-irrigated farming boomed temporarily, but its outcomes for the future were disastrous for both the regional ecology and the economy. Rural poverty was the ultimate outcome, which led the author to investigate its actors and factors through a scientific methodology. He took up the study in 2007, recorded time-series data from 1981 to 2008, and concluded the analysis in early 2012. However, the contents of the study have been thoroughly reviewed and updated for publication as a monograph in 2014.

The author is a university professor with lengthy teaching and research experience. This book derives from his doctoral thesis, which has received great appreciation among academia and field experts working in the line public sector departments and NGOs. Some have recommended presentation of the work as a monograph to the general public in order to influence related social and political strategies. The appreciation of the work was based upon its strong relevancy to the present day's cross-cutting issues of sustainable development, its cohesive conceptual framework, strong methodology, integration of multiple datasets, and plausible findings and conclusions.

The book is divided into seven chapters presenting a thorough context of the regional aqua-agro problems, and setting clear-cut questions, objectives, hypothesis, and significance of the study. It provides a good account of the physical, social, economic, and political features of the study area, which is a vast representative part of Balochistan, to enable a sound environmental perspective for validity and replication of the results in other similar scenarios globally. Besides a detailed literature review outlining what the world intelligentsia currently knows on these fronts, the author has woven an innovative model of how the unsustainable practices in agriculture and water use are inter-effective in the whole socio-institutional and econo-ecological system of human societies. The methodology is well designed and explained. The results are supported by FGDs, sequential satellite images, and field surveys. Standard statistical, image analysis, and cartographic techniques have been used for analysis and presentation of the data. In the back matter of the book, the survey protocols and much additional information about the study area have been annexed.

The study area is located close to the Pakistan-Afghan border. During the period of our field survey, serious risk factors existed here because NATO and US forces were operative against the Taliban regime just across in Afghanistan. However, the author was able to elicit reliable primary data and met all the objectives set for the study. In the front matter of the book, the author has duly acknowledged all contributors in this accomplishment.

I hope the book will be instrumental, both as an educator and as a research template, in several allied fields such as sustainable development, sustainable agriculture and irrigation, land degradation/desertification, rural development, water resource conservation and management, satellite data interpretation, and land-cover analysis and mapping, etc. In the area of public policy formulation, the book contains enough to guide appropriate policy lines and interventions for containment of water crises and sustainable agricultural growth. The rich record of references and additional literature, most with Internet links, is also worthwhile.

Quetta, Pakistan Ahmad Saeed Khattak

# Acknowledgments

I feel profoundly indebted to Professor Dr. Abdul Ghaffar – my research supervisor – who graciously took me as a pupil in the laborious doctoral research. I shall always regard his able guidance, assistance, and sympathies in all stages of this study. The rest of the faculty of the Department of Geography, University of the Punjab, Lahore, also deserve my deepest gratitude for teaching me in the foundation courses, and for their extension of cordial cooperation during my stay at Lahore for the study. I cannot forget my compeers – Mrs. Ibtisam Butt, Mr. Aftab Karim, Mr. Iqbal Abbasi, Mr. Shafqat Anjum, and Mr. Tanveer Zafar Sangra – who greatly contributed to my success by sharing skills in the use of certain sophisticated computer programs and providing pleasing company all the way. Furthermore, I gratefully acknowledge the sincere cooperation of all the concerned chairs in my parent organization – University of Balochistan, Quetta – for a range of support, including the grant of study leave.

My gratitude is also owed to Professor Dr. Amir Khan of Peshawar University, who was a source of valuable advice in the preparation of the research synopsis and the progress onwards. The author expresses his profound regards for Dr. Shahid Ahmad – coordinator of the project entitled 'Supporting the Public Resource Management (SPRM) Balochistan'. I found him a very sincere, devoted, and talented officer and feel myself so lucky to have him at Quetta and to have his instrumental guidance in designing this research.

A major component of this study was the analysis of satellite images. To that end, the contribution of Mr. Haneef-ur-Rahman of SUPARCO, Islamabad, is inexpressible. This gentleman helped me so incredibly that I might not be able to reciprocate in due level. May Allah accept my supplications for a huge reward to him in the Here and the Hereafter. I would also like to acknowledge the valuable assistance of Mr. Ghulam Qadir, Assistant Director of the GSP, Quetta, and Mr. Mohammad Ahmad of WASA (Water and Sanitation Authority), Quetta, in the areas of remote sensing and GIS-based mapping. I must not ignore the friendly reception and help of Mr. Mumtaz Khan Wazir, Chief Engineer (North), Irrigation & Power Department, Balochistan. It was through his reference that I received cooperation in several water resource- and agriculture-related departments and organizations.

I hold sincere wishes for Engineer Kamran Babar of Cameos Consultants, and Mr. Shakeel Azam and Mr. Wajid Ali Shah of the DWRPMD, for the services they rendered to me in collection of the tubewell and meteorological data. Pishin Valley was a very difficult area for me. Therefore, conducting a field survey in this area would barely have been possible had the collaboration of Mr. Naimatullah Tareen not been available. Mr. Tareen is ex-faculty of the Geography Department, Balochistan University, and a native resident of the Pishin Valley. He proved to be an angel of help for me in conducting a field survey in an area that was then strange to me.

I must say that the list of those who contributed somehow in this endeavour is so long that it may be inappropriate to list them all by name. Nevertheless, the author pays tribute to them collectively, especially to the stakeholders and key informants in agriculture and irrigation departments for sparing time out of their busy offices to participate in our survey.

My gracious wife deserves my heartiest wishes for looking after my senile parents, and for being a source of encouragement and solidarity all the way. To conclude, I submit my humbleness to my parents whose prayers made this achievement possible.

Ahmad Saeed Khattak

# Contents

# Abbreviations, Acronyms, and Symbols Used

| | |
|---|---|
| ADB | Asian Development Bank |
| AoI | Area of interest |
| Avg. | Average |
| BKK | Bund Khushdil Khan |
| cfs | Cubic feet per second |
| CGIAR | Consultative Group on International Agricultural Research |
| D. Khanzai | Dab Khanzai |
| DAD(s) | Delay action dam(s) |
| Dr. | Doctor of Philosophy (Ph.D.) |
| DTW | Depth to watertable |
| DWRPMD | Directorate of Water Resources Planning, Monitoring, and Development |
| ETM | Enhanced Thematic Mapper |
| FAO | Food and Agriculture Organisation |
| FGD | Focus group discussion |
| FRP | Flat rate policy |
| Ft. | Feet |
| G. Avg. | Gross average |
| GDP | Gross domestic product |
| GIS | Geographic information system |
| GoB | Government of Balochistan |
| Govt. | Government |
| GSP | Geological Survey of Pakistan |
| HEIS(s) | High efficiency irrigation system(s) |
| IUCN | International Union for Conservation of Nature |
| IWRMP | Integrated water resources management policy |
| KP | Khyber Pakhtoonkhwa |
| Max. | Maximum |
| Mha/or mha | Million hectares |
| Min. | Minimum |
| MSL | Mean sea level |
| MSS | Multi-spectral scanner |

| | |
|---|---|
| N. Malezai | Nale Malezai |
| NGO | Non-Govt. Organization |
| NIPS | National Institute of Population Studies |
| NOCs | No objection certificates |
| O & M | Operation & Maintenance |
| P. evap. | Potential evaporation |
| Pak. | Pakistan |
| PC(s) | Patwar circle(s) |
| PCO | Population Census Organisation |
| ppt. | Precipitation |
| Q. Abdullah | Qilaa Abdullah |
| SOP | Survey of Pakistan |
| SUPARCO | Space & Upper Atmosphere Research Commission |
| TM | Thematic mapper |
| UC(s) | Union council(s) |
| UNCED | United Nations Conference on Environment & Development |
| UNDP | United Nations Development Programme |
| US/ or USA | United States of America |
| USGS | United States Geological Survey |
| Veg. | Vegetables |
| WAPDA | Water and Power Development Authority |
| WCED | World Commission on Environment and Development |
| $<$ | Less than |
| $>$ | Greater than |
| $\leq$ | Less than or equal to |
| $\geq$ | Greater than or equal to |

# List of Appendices

# List of Figures

# List of Maps

# List of Tables

# Chapter 1
# Introduction

**Abstract** Problems often emerge through broad perspectives. For understanding and investigation, tracking them by the perspectives is helpful. This chapter introduces the problem of tubewell farming versus aquifer potential through an account of its background issues. Since the problem has countrywide implications, it has been tracked from national through provincial contexts and finally short-handed at local level scenario for scientific analysis. Further, this chapter settles the research questions, objectives, and hypothesis to be pursued in the study. Besides establishing its operational scope, it has provided a niche to this research in the domain of geographic research literature – to be considered specifically as one of the agriculture and water resources management studies.

**Keywords** Balochistan • Pishin Valley • Agriculture • Groundwater • Tubewell irrigation

## 1.1 The Context

Agricultural development, increasing human and livestock populations, and drought have degraded natural resources in the arid and semi-arid regions of the world (Amissah-Arthur et al. 2000). In particular, access to freshwater resources is one of the major challenges being faced by humans and livestock for their survival. Therefore, judicious use and management of the scarce water resources has become essential to achieving food security at all levels – from global through local (Tiwari and Dinar 2002).

As Pakistan in primarily an arid country, irrigated agriculture is predominantly practiced. However, high fluctuations in precipitation regime, limited water storage capacity, and a poor use of the available water resources consistently result in much slower growth of irrigated agriculture as compared with the ever-increasing needs of the population. During the recent drought (lasting from 1998 to 2004), production of cotton bales was down by 2.1 million, while production of wheat, sugarcane,

A.S. Khattak, *Mutual Sustainability of Tubewell Farming and Aquifers: Perspectives from Balochistan, Pakistan*, Advances in Asian Human-Environmental Research,
DOI 10.1007/978-3-319-02804-0_1, 

and rice was lower by 4.1, 7.6, and 1.2 million tons, respectively, during 2000–2001 as compared with pre- drought levels (Alam and Naqvi 2003). Agriculture contributes 24.5 % to the GDP. It employs more than 45 % of the country's total labor force and directly or indirectly supports the living of about 68 % of the population. Further, it contributes 65 % to the total national export earnings derived from raw and processed agricultural commodities (Alam and Naqvi 2003).

While the transition to an urban and industrial economy is progressing in Pakistan, it appears that agriculture will remain central in providing a living to a large number of people. Good management of water resources is key to improving agricultural productivity and creation of jobs. The persistence of the water economy over the last several decades has not been based on interventions by the state; rather, it has been privately run by millions of farmers and industries through the tapping of groundwater. It is now obvious that this era of 'productive anarchy' is close to an end, because groundwater is now being over-tapped and exhausted in many parts of the country, including the Indus Basin and other areas such as the Uplands Balochistan. This scenario poses two very serious challenges to the state. First, the surface water supply systems are resuming their previously high level of importance and are, therefore, required to be managed more effectively than ever before. Second, both the quantity and quality of groundwater will have to be managed much more aggressively than in the past. The role of water as a social, economic, and ecological good should be reflected in demand-management mechanisms implemented through resource assessment, and water conservation and reuse strategies (UNCED 2002).

### 1.1.1 Agricultural Production Constraints: National Perspective

Population is increasing most rapidly in those parts of the world where the food supply is least adequate and where incomes are the lowest (Mosher undated). This is equally true in the case of Pakistan. The country is presently in the grip of a severely intense population explosion, and has experienced greater population growth than other developing countries. The population of Pakistan is increasing at an annual rate of 2.61 %, hence the gap between demand and supply of agricultural products is widening year after year. After the creation of Pakistan in 1947, the first population census, in 1951, enumerated about 33.74 million people. Nearly half a century later, Pakistan conducted its fifth and most recent census in 1998, enumerating a population of 132.35 million, 67.5 % of which lived in rural areas (PCO 1998). According to projected estimates, the population of Pakistan was 161.86 million in the year 2010; it would be 175.65 million, 189.42 million, and 202.11 million by the years 2015, 2020, and 2025, respectively (NIPS 2006). Pakistan is the world's seventh biggest country in terms of population, but in terms of area, it occupies only 0.67 % of the world's land (Nawaz 2004). Of the total 79.61 mha land area of Pakistan, 20.69 mha is under cultivation, of which only 5.34 mha can sustain intensive agriculture (Ahmad et al. 1998).

The agriculture framework of Pakistan is supported mainly by the crop sector whose contribution to agricultural GDP exceeds all other sectors of the agro-economy. In the years 2000–01, the crop sector contributed 40 % to agricultural GDP as compared with 38 % from livestock and 5 % each from the fishery and forestry sectors (Alam and Naqvi 2003). However, the cropped area is not growing at pace with the country's population, which has put enormous pressure on farmers to obtain greater yield of food and fiber per unit of land. Since our industrial base is dominantly agrarian, rapid agricultural growth is also required to stimulate the pace of industrial growth, thus setting into motion a mutually reinforcing process of sustained economic development (Alam and Naqvi 2003). It is thus evident that the well-being of the vast majority of the population is critically dependent upon efficient and sustainable utilization of the agricultural resources of the country.

### *1.1.2 Land and Water Constraints: National Perspective*

Pakistan has a diverse topography. Its northeast, north, and northwestern fringes are mostly mountainous, and the total area of the country is 7,96,096 sq. km (PCO 1998). The cultivable area is 35.4 mha, forest land 3.5 mha, cultivable waste 8.6 mha, cultivated area 22 mha, waterlogged- and salinity-affected area inside the Indus Basin is 6.8 mha, while outside the Indus Basin it is 6.3 mha. The desert lands of the country include Chaghi-Kharan (Balochistan), Thar (Sindh), and Cholistan and Thal (Punjab) (Alam and Naqvi 2003). Despite commendable progress in the agricultural sector of the country's economy, several weaknesses also exist in this sector. The most fundamental constraint is water scarcity, which hampers adequate and reliable irrigation supplies and limits further expansion of irrigated agriculture (Alam and Naqvi 2003). Other constraining factors include widespread occurrence of waterlogging and salinity, floods, soil erosion, low yield per acre, and primitive cultivation methods. Pakistan's climate is largely dry, hence the truly profitable agriculture could be that which is irrigated. Unfortunately, however, extreme weather conditions (either drought or flood), meager storage capacity, and traditional irrigation practices culminate in the water resource acting as a constraint to agricultural growth rather than an opportunity.

Over the past four to five decades, the exploitation of groundwater, mostly by private farmers, has brought tremendous agricultural development. But now there are clear indications that, owing to over-exploitation of groundwater, the current pace of development is not sustainable. In the Uplands zone of Balochistan (where the study area lies), farmers are pumping out water from depths of hundreds of meters. Further, problems with groundwater quality are emerging. These facts make it imperative to develop policies and approaches for bringing water withdrawals into balance with recharge. This may be a difficult process as it would require some harsh political and administrative actions by government, and sacrifices by the organized and influential resource users.

### 1.1.3 *Tubewell Farming and Ecology: Provincial Perspective*

The Balochistan Province of Pakistan occupies its global position between longitudes 60° 50′ to 70° 5′ E and latitudes 24° 50′ to 32° 10′ N. It occupies the south-western part of the country, and has a 770 km long coastline in the south along the Arabian Sea (Syed 2004). Among the four provinces of Pakistan, Balochistan is the largest in terms of area (total geographical area 34.72 mha/or 3,47,190 km$^2$) and covers about 43.6 % of the country's map (GoB 1992–93, 2013). But its population of 6.5 million (as per the latest census of 1998) is only 5 % of the country's total, of which 76.1 % is rural (GoB 2013; Nawaz 2004; PCO 1998).

Agriculture is the mainstay of the economy; it contributes around 30 % of the province's GDP (Ahmad and Khan 2007; ADB et al. 2007) and employs 67 % of its labor force (Syed 2004). According to agricultural statistics from 1996 to 1997, 58.6 % (20.35 mha) of the total geographical area of Balochistan is not available for cultivation: forests cover 5 % (1.87 mha); cultivable waste accounts for 25.1 % (8.72 mha); and the area under cultivation is only 11.3 % (3.79 mha) (GoB 2013). It is also important to note that 60.01 % of the total area under cultivation is current fallow (6.5 % of the total land area); only the remaining approximately 40 % is net sown (4.8 % of the total land area). Irrigated agriculture has a significant share in the economy of the province; almost 50–54 % of the labor force is directly or indirectly engaged in this sector (Saeed 2006). Using statistics from 2005 to 2006, a study estimated that around two-third (1.28 mha) of the cultivated area is irrigated and the rest (0.65 mha) is under sailaba (flash flood irrigated) and khushkaba (rainfed) farming (Mirza and Ahmad 2008). Of the total irrigated area, the area served by the Indus Basin irrigation system is 46 %, followed by tubewell irrigation system, with a ratio of 36 %. The karezes, springs, river diversions, and infiltration galleries provide irrigation to 10.93 % of the area collectively, while open wells serve 6.32 % of the total area under irrigation (Saeed 2006).

It has been concluded that Balochistan's agricultural sector grew phenomenally during the 1980s. But afterward, the natural resource base of the province has faced the threat of losing its economic potential due to deteriorating ecological conditions. Inefficient water use, wastage of surface water, and indiscriminate exploitation of groundwater, coupled with historic water scarcity, aggravated the situation further, making water management complex in Balochistan (GoB 2004). Moreover, the acute drought during 1997–2004 has badly impacted on the availability of water and livelihood of rural communities. While it is imperative that these resources be used most judiciously to ensure sustainable agricultural development and productivity levels, to date, there has been no comprehensive water policy in the province (GoB 2004).

The climate of Balochistan is predominantly arid to semi-arid. Mean annual rainfall ranges from <50 mm in the southern to 400 mm in the northern parts of the province. Due to aridity, irrigation is central to achieving and sustaining food security in the province. During 2005–2006, around 1.93 million hectares of land was under cultivation. Of this, 1.28 million hectares (66.32 %) are irrigated and the rest are rain-fed (Ahmad 2006a, b, c). Around 1.36 million hectares are forested, which largely have sparse vegetation. In addition, there are 4 million hectares of wastelands which could be cultivated if water were available (Saeed and Ahmad 2008; GoB 2006a, b).

The projected population of the province for 2006 was around 8.3 million and was growing faster (2.2 % per annum) than the national growth rate (1.66 % per annum) (NIPS 2006). According to the projected estimates by the PCO of Pakistan on the basis of the 1998 census, the population of Balochistan Province was 7.74 million in 2005, and 8.57 million in 2010; and it was projected to be 9.46 million, 10.34 million, and 11.1 million during the years 2015, 2020, and 2025, respectively (PCO 1998). Agricultural statistics and other reports indicate that the gap between demand and supply of agro-stuff is increasing. Although Balochistan still exports fruits and vegetables to other provinces of Pakistan and globally, it has now become a net importer of several food commodities. This is primarily due to the low expansion rate of cropped areas and small farm productivity compared with the population growth rate.

There is no dearth of cultivable land in Balochistan. Water is indeed a resource at premium, which adds to the value of land from almost nil to a very expensive commodity in areas where ample ground- or surface water is available. At the time of Pakistan's independence in 1947, agriculture of the province was limited to areas that had perennial water supplies, either through canals, karezes and springs, or through dug wells. However, agriculture expanded with water supplied from the Pat Feeder Canal and tubewells. The Pat Feeder Canal was constructed between 1963 and 1969 in Kachi Plain as part of the Guddu Barrage Project to irrigate Naseerabad and Jaafarabad Districts of Balochistan. Tubewells flourished during the 1970s and 1980s with expansion of electrification. Consequently, the total cropped area of the province, which was only 636,083 ha during 1984–85, grew within a decade to 906,691 ha during 1994–95 (GoB 1994–95). However, since then, the pace of increase in cropped areas has generally declined; for example, the total cropped area during the year 2000–01 was 840,132 ha (GoB 2000–01).

Although modern tubewell technology has enabled farmers to abstract water from much deeper levels to enhance tubewell farming, there are serious concerns regarding the sustainability of irrigated agriculture, especially the tubewell irrigated type. In fact, groundwater constitutes only around 4 % of the total water resource available to the province (Ahmad 2006a, b, c). The problem of watertable decline is extremely serious in the highlands of Balochistan, which account for approximately 53 % of the province (Nawaz 2004). For instance, in parts of the Pishin-Lora Basin (Appendix D), the decrease of the watertable is quite rapid, ranging from 8 to 10 ft per annum (Nawaz 2004). Consequently, karezes, springs, and flows in seasonal streams are dwindling and desertification has surfaced in many previously productive lands.

Among all the sub-sectors of water consumption (agriculture, humans, livestock, industry, mining, and nature), agriculture consumes around 97.5 % of the total water resources available in the province (Ahmad 2006a, b, c). The number of tubewells has increased dramatically with the introduction of the national electricity grid system in the 1970s, as well as a program subsidizing electricity for agricultural tubewells (Nawaz 2004). There are 25,734 operating tubewells in the province, 14,363 of which are electric operated and 11,371 are diesel operated (GoB 2004). The annual growth rate of tubewells over the last 17 years was around 6.7 and 7.4 % for electric and diesel operated tubewells, respectively. Despite the lowering of the watertable and groundwater mining at an alarming rate, installation of additional tubewells is continuing unchecked (GoB 2004).

The depletion of groundwater has adversely affected water quality in some of the river basins (Ahmad 2006a, b, c) of Balochistan. For instance, in parts of the Pishin Sub-Basin, the groundwater has become unsuitable for both domestic and irrigation use due to its brackishness. Consequently, agricultural development is under serious threat in those areas. Besides the low expansion rate, intensification and the economic efficiency of irrigated agriculture is also questionable (Ahmed and Ahmad 2007). The IWRMP for Balochistan, 2006, has therefore highlighted the need to address all sources of water, and all sub-sectors of water use, with special focus on managing the scarce groundwater resources for future generations. In particular, the Pishin-Lora Basin, which is the most overdrawn basin in the province, is the focus of Government interventions (Riaz et al. 2008; GoB 2006a).

The Pishin-Lora Basin lies in the northern part of Balochistan Province between latitudes 28° 43′ and 31° N and longitudes 66° 12′ and 67° 43′E (Appendix D). Population of the basin was 2.18 million in 1998; projected population for 2007 is 2.71 million. The basin catchment area is 18,133 km$^2$, out of which alluvium occupies about 7,873 km$^2$ (Riaz et al. 2008). It is further subdivided into eleven sub-basins: Pishin, Kuchlagh, Quetta, Kolpur Sardar Khel, Mastung, Shirinab, Patki, Shahnawaz, Mangochar, Kalat, and Kapoto. The study area is part of the Pishin sub-basin. The soils are of alluvial origin and suitable for growing high-value crops and deciduous fruits if irrigation were available.

## 1.2 Problem Definition

The study area is Pishin Valley, which includes the piedmont and plains of the Pishin sub-basin; that in turn is part of the large Pishin River basin. In many parts of the Valley there is phenomenal and rapid growth of irrigation tubewells. Groundwater is being abstracted using high-powered pumps without any practical checks. Record shows that some regulatory restrictions exist on paper; however, their practical implementation is not seen on the ground. Another reason for the tubewell boom may be the so-called ‘flat rate policy’, under which the farmers pay a fixed amount of Rs. 4,000 per tubewell per month, irrespective of the actual consumption of electricity. The District Water Committee (DWC), which is responsible for sanctioning NOCs to farmers for the drilling of tubewells on set terms and conditions, appears unable to enforce groundwater licensing regulations (postulated in the Groundwater

Rights Administration Ordinance, 1978) or any penalty to defaulters. The law and order situation and lack of political will are obviously the major reasons for institutional weaknesses.

Apparently, the subsidy on electricity tariff, via which the farmers of the province pay only around 9 % of the total electricity bill, resulted in wasteful use of scarce groundwater resources. Violent opposition from the 'Tubewell Farmers' Action Committee' is preventing the Government from phasing out the subsidy (Ahmad 2006a, b, c). Political will is essential to address the issue of the tubewell subsidy. Indeed, the provincial Government has made the decision to cap the number of tubewells eligible for electricity subsidy to 15,206 and to fix the GoB's share in the total required subsidy to a lump sum of Rs. 2.0 billion (accounting for 28.4 % of current actual electricity consumption). These decisions reflect a realization among provincial Government circles that the subsidy issue should be addressed properly in the future, but a new vision and revised strategy would be required to resolve it in a socially and politically acceptable manner (Ahmad 2007).

The indiscriminate exploitation of groundwater has resulted in rapid decline of the watertable and mining of groundwater resources. As a result, a large number of the tubewells have dried out, causing farmers to lose huge amounts of money. In addition, the water yield of the tubewells also gradually diminishes, forcing farmers to own more than one tubewell at a time. Further, in some parts of the Valley, degradation of the water quality has also emerged. Due to this scenario, desertification in agricultural lands has begun (Appendix E) and tubewell-irrigated agriculture has become less economically productive, particularly for the provincial economy as a whole (the actual irrigation cost is too much higher than what farmers pay). On the face of these facts, the most hit and vulnerable are the small and resource-poor farmers, who can neither afford to have their own tubewells nor get sanctioned for them the subsidy in electricity tariff. Thus, with the continuous deterioration of the local aquifer, which is the only source available for irrigation, the social sustainability of irrigated farming has also become at stake.

It seems essential, therefore, to strictly enforce the Groundwater Ordinance of 1978, and develop new mechanisms for effective institutional support. Besides reforming the issue of subsidy, dissemination of efficient pumping and irrigation technologies and training to farmers could be key potential measures for managing the demand. These have to be accompanied by recharging the groundwater by expanding the sailaba farming through diversion of around 12 billion m$^3$ of floodwater that is being wasted and that causes damage to infrastructure during flood years – a good example is the flood of 2007 (Ahmad 2007) . Thus, the outcome of the proposed research would be the answer to "how can the available scarce surface and groundwater resources in Pishin Valley be managed to sustain population and agricultural growth on a longer-term basis, and how could future inter-generational issues on water use be avoided? Further, the groundwater has to be treated as a vital but scarce resource in the Valley, and future generations must not be deprived of its use through unsustainable exploitation practices."

## 1.3 Objectives of the Study

The general objective of this research work is to conduct a spatio-temporal analysis of the development of tubewell irrigated agriculture in Pishin Valley, and its relationship with the potential of the aquifer, to formulate a revised strategy by making 'groundwater as business for everyone'. The specific objectives are to:

1. Assess the changes in horizontal (expansion) and vertical (intensification) development in agricultural land use during the past three decades using field-generated data, as well as the modern techniques of remote sensing.
2. Find out water scarcity-driven changes in the cropping pattern.
3. Evaluate the cost-benefit trend of tubewell irrigated agriculture to show how its economic viability is at stake in view of the deteriorating aquifer.
4. Assess the limits and rates of watertable drawdown, mining of the aquifer, and qualitative degradation of groundwater in the Valley.
5. Examine the prevailing agriculture support strategies of the Government from the standpoint of their impacts on sustainability of the groundwater economy of the region.
6. Suggest a new vision and revised strategy for addressing the issue of groundwater mining in a more comprehensive way by adequately educating and equipping all stakeholders.

## 1.4 Research Questions

To achieve the objectives of the study, the following questions have been framed to find answers to via this study:

- What are the major inter-effects of the developments in tubewell irrigated agriculture and the aquifer potential in Pishin Valley considering the regional ecological, social, and policy perspectives?
- How have public policies and development programs contributed to the severity of the problem of watertable depletion and mining of groundwater?
- What are the significant and sustainable interventions introduced by public sector departments and civil society organizations in Pishin Valley for managing the groundwater resources?
- Why could public and civil society interventions not contribute significantly to averting aquifer deterioration?
- How can a new vision and revised strategy be developed to bring on board social, political, and policy support for addressing the issue of groundwater mining in a comprehensive and effective way?

## 1.5 Hypothesis

This research aims to test and validate the hypothesis given hereunder. The test results of this hypothesis can be generalized to almost the whole of Balochistan Province as it bears almost uniform conditions as the issues under study. The hypothesis states:

> The progress of tubewell farming and the aquifer potential are affecting each other in an unsustainable manner in Pishin Valley under the prevailing ecological conditions, socio-political considerations, and policy framework.

## 1.6 Scope of the Study

Agricultural development is a broad concept. Operationally, here the term is used to mean spread of irrigated agriculture to previously rain-fed lands or cultivable-waste; farm intensification in terms of multiple cropping and intercropping; and rise in farm profitability in financial terms. By aquifer potential we mean the watertable stability (resource quantity) at cheaply exploitable depth levels, and the suitability of water quality for multiple uses – domestic and irrigation. In essence, we are looking at the sustainability issues of the prevailing irrigation and agricultural practices in the context of their impacts on local ecology, social equity, and cost effectiveness. The study is reviewing the local social characteristics and regional policy framework as they have crucial bearing on the problem under study.

## 1.7 Significance of the Study

At this juncture, when 'rural' is taking on new dimensions in the face of long-standing challenges, agriculture has become a prime topic of interest. Sustainability of agriculture is an issue that excites people all over the world. In Balochistan Province of Pakistan, as well as in most of the rest of the country, population expansion, coupled with rising desires for high living standards, has increased pressure on the already marginal and fragile natural resources to the extent of endangering their sustainability for coming generations. Agriculture is under serious stress due to dry climate and lack of irrigation opportunities. Consequences are rising rural poverty, underemployment, and food insecurity. To meet the growing demands for food, fiber, and cash income, expansion and intensification of irrigated agriculture, and its cost effectiveness are the solutions. However, there is an educated perception that most of the existing agriculture-related policies and practices in Balochistan are inconsistent with these objectives.

A study of this nature, which integrates the two-way flow of impacts between nature (aquifer) and a human enterprise (agriculture), is one of remarkable contribution to the field of geography. It is subscribing to the literature of sustainable development, agricultural geography, water resource management, and arid zone ecology (a subset of environmental geography), whereas, all these fields have not yet attracted enough attention among geographers in Pakistan. The book presents and empirically applies a number of methods for research synthesis based on spatial and temporal analytical principles. Not only could the results of this study currently be incorporated in the Integrated Water Resource Management Policy of Balochistan, they would also equally be applicable in other parts of the world with similar challenges and opportunities. In addition, this study could enhance public awareness about the potential problems associated with the unsustainable exploitation of the limited natural resource bases of the relevant regions. Enhanced public awareness regarding the importance of irrigation resources in worldwide agricultural economies could lead to their greater support for implementation of conservation strategies.

Although, the study has been carried in the peculiar physical, social, and political scenario of Pishin Valley, Balochistan, it presents a conceptual framework within which certain hypotheses about sustainable development and sustainable use of natural resources and their indicators may be tested, and surrogate generalizations may be formulated from their results. Because depletion of natural resources and desertification of groundwater irrigated lands are emerging worldwide hazards, it is rational to expect the findings of this study to also be replicable in other countries, especially those with similar environmental set ups.

## 1.8 Limitations of the Study

The study at hand is a time series analysis to cover a span of 28 years of development and exploitation in the past. There is nothing better than a well recorded complete data for a study such as this. However, we admit two major limitations in our work:

1. The foremost is that most of the farmers of the study area were uneducated simple folk and had not kept any record of the historical facts and figures of their tubewells, such as their depth 25 years ago, particularly when many previous tubewells had dried up due to watertable fall and been replaced by others. Similarly, they had no hard and fast facts regarding the previous costs of irrigation or farm-income statistics. Thus, the data collected for anything other than the present (year 2008), presumably older than 8 years, may not be completely accurate. However, we believe that our strategy of extending the data temporal intervals to 10 years, instead of 4 years, like in the last two temporal intervals, proved helpful in enabling respondents to recall their best memories to provide reliable approximations. Of course, we felt some cases exaggerated their miseries to a worst-case scenario; however, we had to listen and register them as it was

part of the strategy. Leading respondents to a particular answer is not advisable. It is also worth mentioning that data of the nature we required is not usually recorded by public departments in our part of the world; hence, we were dependent on farmers' output.

2. While designing the project, we anticipated the data constraints discussed above. Therefore, we complemented the farmers' narratives with an eye in the sky – satellite images – as this provides exemplary reliability in diagnosing land-cover changes and was therefore the best possible solution. This is all well and good; however, the limitation is that we could not produce satellite data for 1981 or thereabouts, because until then no Pakistan Government agency had that information. Further, the satellite data itself had many limitations by that time. Since 1981 was the starting year for analysis in this project, it would have been better had we been able to provide the scenario for that year via such an authentic source. To an extent, satellite data for 2008 was also justified as it was the closing point for our analysis; however, because of its high cost, we used the 2005 imagery as representative for the whole of that 4-year period (2005–2008) because no extensive change in land cover was expected in the intervening years.

## 1.9 Organization of the Book

The subject matter of the book is provided in seven chapters:

Chapter 1 introduces the basic ideas of this study and provides context for the research problem. It also sets the objectives, questions, hypothesis, scope, and limitations of the study, and highlights its significance for the society at large.

Chapter 2 lays down the conceptual framework for the present study and recognizes related ideas and findings from previous literature.

Chapter 3 describes the study area at length. It includes physical (e.g. climate, relief, soil) and socio-economic aspects (e.g. population, education, occupations, tenure system) of the area.

Chapter 4 explains the applied research methodology; and reveals the types and sources of data, sampling techniques, data collection, and analysis techniques.

Chapter 5 provides results and discussions based upon the field survey and satellite data regarding socio-economic features of the farmers, agricultural land-use dynamics, cropping pattern issues, and farm income.

Chapter 6 provides results and discussions based upon the data collected through two separate field survey protocols regarding aquifer issues and the role of government in agriculture and water resource management.

Chapter 7 draws conclusions, summarizes key findings, and puts forth policy suggestions for reforms.

## 1.10 Summary

This chapter introduced the research problem and discussed its background at various levels. It highlighted the need for careful management of water resources as their sustainability is vital for agriculture, which sustains humanity in many ways. The chapter showed that Pakistan's economy is agriculture-based, as it contributes 24.5 % of the GDP and employs around 45 % of the workforce. However, while the country's population is rapidly increasing, cropped area is growing at a slower pace, mainly due to climatic constraints and irrigation limitations. In the provincial context, agriculture is even more important, contributing around 52 % of the province's GDP and employing 65 % of its labor force. The chapter revealed that Balochistan's agriculture was at its peak during the 1980s, but has since been consistently under threat in terms of its multidimensional sustainability. Particularly in Pishin Valley, where tubewell farming is the dominant productive land use, the inefficient irrigation culture and the inappropriate cropping pattern, etc., have adversely affected the natural resource base, which in turn has implications for the social and economic viability of tubewell farming. Moreover, this chapter has oriented the whole study by establishing questions, objectives, hypothesis, etc.

## References

Ahmad S (2006a) Integrated water resources management in Balochistan, vol IV. ADB, Pakistan Resident Mission, Islamabad, TA-4230 (Pak)

Ahmad S (2006b) Issues restricting capping of tubewell subsidy and strategy for introducing the smart subsidy in Balochistan. TA-4560 (Pak), policy briefings, vol 2, no. 1. Govt. of Balochistan, ADB, and Royal Netherlands Govt, Quetta

Ahmad S (2006c) Balochistan's water sector: issues and opportunities. Balochistan economic report, prepared by the Joint Economic Mission to Balochistan, GoB, ADB, and World Bank, Quetta

Ahmad S (2007) New vision and strategy for managing water and energy use in tubewell irrigated agriculture of Balochistan. TA-4560 (Pak), water for Balochistan policy briefings, vol 3, no. 2. Govt. of Balochistan, ADB, and Royal Netherlands Govt, Quetta

Ahmad I, Ahmad S (2007) Water productivity and economic efficiency of tubewell irrigated farms in Balochistan: issues and policy reforms. TA-4560 (Pak), water for Balochistan policy briefings, vol 3, no. 11, 2007. Govt. of Balochistan, ADB, and Royal Netherlands Govt, Quetta

Ahmad S, Khan AG (2007) Sailaba and khushkaba farming systems of Balochistan – policy support for changing landuse and to avoid infrastructure damages caused by flash floods. Water for Balochistan, policy briefings, vol 3. Govt. of Balochistan, ADB, and Royal Netherlands Govt, Quetta, Accessed at: http://www.spate-irrigation.org/wordpress/wp-content/uploads/2011/06/PolicyBriefing20.pdf. Accessed 26 Jan 2014

Ahmad B, Ahmad M, Gill ZA (1998) Restoration of soil health for achieving sustainable growth in agriculture. Pakistan Institute of Development Economics, vol 37. The Pakistan Development Review, Islamabad

Alam SM, Naqvi MH (2003) Pakistan agriculture – 2003. http://pakistaneconomist.com/pagesearch/Search-Engine2003/S.E155.asp Accessed 15 Jun 2007

Amissah-arthur A, Mougenot B, Loireau M (2000) Assessing farmland dynamics and land degradation on a Sahelian landscape using Remotely sensed and socio-economic data. Int J Geogr Inform Sci 14(6):583–599

ADB (Asian Development Bank) et al (2007) Supporting public resource management in Balochistan: basin-wise water resources availability and use. TA-4560 (Pak), Final report, pp 6, 18. Prepared by Halcrow and Cameos Consultants for Govt. of Balochistan, ADB, and Royal Netherlands Govt., Quetta

GoB (Govt. of Balochistan) (1992–93) Development statistics of Balochistan. Bureau of Statistics, Planning & Development Department, Quetta

GoB (1994–95, 2000–01) Agricultural statistics of Balochistan, statistics wing, Directorate General of Agricultural Department, Balochistan, Quetta

GoB (2004) Integrated water resources management policy for Balochistan. Final draft report, component # 3. Balochistan Resource Management Programme, Quetta

GoB (2006a) Integrated water resources management policy for Balochistan. Department of Irrigation and Power, Govt. of Balochistan, Quetta

GoB (2006b) Agriculture statistics of Balochistan. Directorate of Crop Reporting Services, Department of Agriculture, Quetta

GoB (2013) Irrigation and power. Govt. of Balochistan official website. Accessed at: http://www.balochistan.gov.pk/index.php?option=com_content&view=category&id=1093&Itemid=65. Accessed 17 Jul 2013

Mirza HH, Ahmad S (2008) Restructuring and strengthening of water resources planning, Development and Monitoring Directorate of Irrigation and Power Department, Balochistan. Water for Balochistan policy briefings, vol 4, no. 1, TA-4560 (Pak). Govt. of Balochistan, ADB, and Royal Netherlands Govt, Quetta, Accessed at: http://watsan.moenv.gov.pk/Documents/Publications/Water%20for%20Balochistan/4.Restructuring%20and%20Strengthening.pdf. Accessed 17 July 2013

Mosher AT (undated) World food problems and prospects. Accessed at: http://ageconsearch.umn.edu/bitstream/17572/1/ar670021.pdf. Accessed 13 Jul 2008

Nawaz K (2004) Community participation and water management in Balochistan, Pakistan. Accessed at: http://archive.unu.edu/env/workshops/Aleppo/06%20-%20Nawaz.doc Accessed 16 May 2008

NIPS (National Institute of Population Studies) (2006) Population projections of Pakistan. National Institute of Population Studies, Govt. of Pakistan, Islamabad

PCO (Population Census Organization) (1998) Census report of Pakistan. Statistics Division, Government of Pakistan, Islamabad

Riaz M, Sani B, Babar K, Ahmad S (2008) Potential recharge zones of over-drawn river basins of Balochistan, Pakistan. TA-4560 (PAK), policy briefings, vol 4, no. 8. Govt. of Balochistan, ADB, and Royal Govt. of Netherlands, Quetta, Accessed at: http://watsan.moenv.gov.pk/Documents/Publications/Water%20for%20Balochistan/3.Potential%20Recharge%20Zones%20of%20over%20drawn.pdf. Accessed 15 July 2013

Saeed M (2006) Promising crops and water efficient cropping patterns for irrigated farming systems of Balochistan, Pakistan. TA-4560 (Pak). Summary report. Govt. of Balochistan, ADB, and Royal Netherlands Govt, Quetta

Saeed M, Ahmad S (2008) Promising crops and water efficient cropping patterns for irrigated farming systems of Balochistan, Pakistan – key issues and policy reforms, TA-4560 (PAK), vol 4, no. 2. Asian Development Bank, Quetta

Syed FH (2004) Economic development of Balochistan. Shabir-Ud-Din, Karachi

Tiwari D, Dinar A (2002) Balancing future food demand and water supply: the role of economic incentives in irrigated agriculture. Q J Int Agric 41(1/2):77–97

UNCED (2002) Conservation and management of resources for development. UN Conference on Environment and Development, Johannesburg Summit, 26th August to 4th September, Agenda 21, Chapters 14–18

# Chapter 2
# Conceptual Framework and Literature Review

**Abstract** While 'conceptual framework' means a researcher's own perceptions about the scope and structure of a problem, the literature review provides others' ideas and work in areas close to that under study. With such a philosophy in mind, this chapter first constructs the author's own thinking as to how the problem in question has originated, disseminated, and evolved; and how it can be resolved for sustainability if turned around. It also draws support from other like-minded models proposed in the fields of sustainable development. Next, the literature review thoroughly, but not exhaustively, scans the depth and breadth of the existing body of knowledge in the areas of sustainable agriculture, irrigation, and water resources, especially groundwater management. This chapter serves to elucidate the concept of sustainable development. It indicates the measurable indicators of sustainability in agriculture and irrigation systems, thus paving the way for designing the data tools at a subsequent stage. With the intention of giving due proportionate space to every significant prong of the aqua-agro issue, the whole conceptual discussion is organized under three sub-topics: (1) the general concepts of sustainability and sustainable development; (2) the focused area of sustainable agricultural development; and (3) the sustainable use of groundwater as an irrigation resource. To highlight the gap in indigenous research literature, the literature review is also organized in two levels of literature: international and national.

**Keywords** Sustainability • Sustainable development • Groundwater management • Over-exploitation • Efficient irrigation

## 2.1 Conceptual Framework

This study derives its basic methodology from the general concepts of 'sustainable development' and 'systems theory'. The focus is on groundwater as an inevitable irrigation resource, and on agriculture as the principal source of rural livelihood.

A.S. Khattak, *Mutual Sustainability of Tubewell Farming and Aquifers: Perspectives from Balochistan, Pakistan*, Advances in Asian Human-Environmental Research, DOI 10.1007/978-3-319-02804-0_2, 

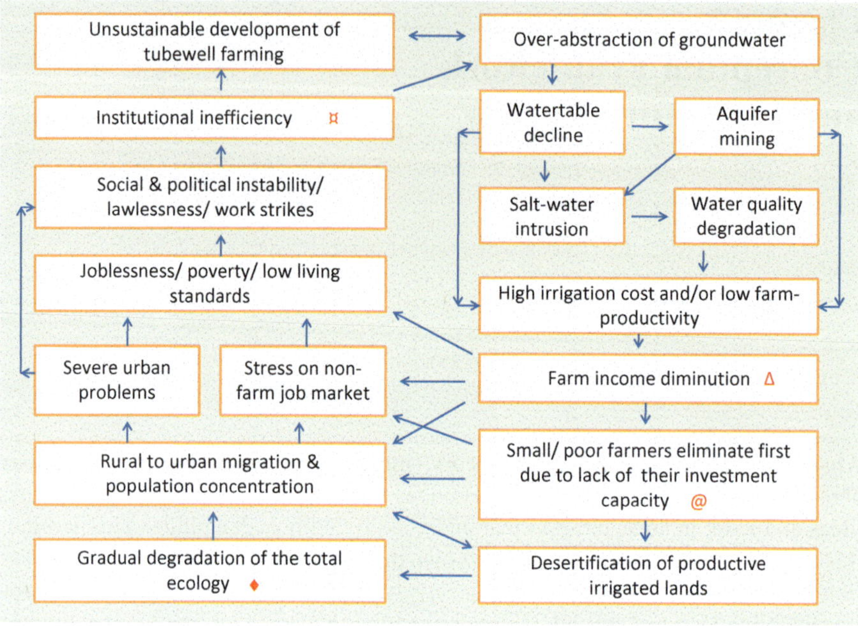


**Fig. 2.1** Conceptual framework of the study (Source: Saeed 2012)

Sustainable management of these two inter-affecting phenomena can achieve the goal of sustainable rural development at the local level, and forward into secure regional, national, and even global prosperity. Since the study site is poor in terms of quantity, continuity, and reliability of data, we could not use any specific analytical model in a harder sense. However, our analysis and assessment acquired basic knowledge from one of the most integrated modeling frameworks, known as 'System for Environmental and Agricultural Modeling; Linking European Science and Society -Integrated Framework (SEAMLESS -IF)'. This model is basically a tool for mutual impact assessment of the various components within an agricultural system, such as those linked up in Fig. 2.1.

### *2.1.1 Sustainability and Sustainable Development*

'Systems theory' is an interdisciplinary field that studies the nature of complex systems. In the most general sense, a system means a configuration of parts connected and joined together by a web of effective relationships. Systems theory is typically used to guide decisions on issues, such as resource use and conservation policies, research and development in technology, rural and urban development, etc.

Systems theory assumes that, no matter how complex and diverse the world is, it will always be possible to find various types of organizations within it. Following the systems theory, the basic idea of sustainability appears straightforward; as a sustainable system is simply a system that is able to survive or persist (Costanza and Patten 1995). In extension, a system contributes to a sustainable development if its relationships to other co-existing systems, including the higher-level system in which it may be embedded, do not hinder their existence. A system's performance in response to impacts is commonly evaluated on the basis of attributes, which are used to derive indicators. The selection of appropriate attributes depends on the specific problem and the way the system is described. Some approaches base selection of indicators and sustainability assessment on understanding of the cycle of human-induced environmental change. A widely used concept in this respect is the DPSIR (Driving force – Pressure – State – Impact – Response) framework. It has been adopted by a number of impact-assessment models, e.g. IMAGE 2.0 (Alcamo et al. 1994a, b).

Recent studies stress the importance of scenario analysis as a powerful tool for integrating knowledge, scanning the future in an organized way and internalizing human choice into the science of sustainability (Swart et al. 2004). According to Schoute et al. (1995), a scenario is a description of the current situation, of a possible or desirable future state, as well as of the series of events that could lead from the current to the future state. Scenarios do not aim at predicting, but rather at exploring, the future (van Ittersum et al. 1998) through possible future states. In a policy decision context, scenarios allow policy makers to anticipate and assess the risks involved in different options and to identify alternative courses of action (Malafant and Fordham 1997).

Sustainability integrates three main goals – environmental health, economic profitability, and social equity. There are many alternative definitions of sustainability and sustainable development. For instance, the WCED, also known as the Brundtland Commission, has defined sustainable development as "the process of change in which the exploitation of resources, the direction of investments, the orientation of technological development, and institutional change are all in harmony and endorse both the current and future potential to meet human needs and operations" (WCED 1987). A year later, the FAO of the United Nations in Rome defined sustainable development as "the management and conservation of the natural resource base and the orientation of technological and institutional change in such a manner as to ensure the attainment and continued satisfaction of human needs for present and future generations. Such sustainable development in the agriculture, forestry and fisheries sectors conserves land, water, plant and animal genetic resources, is environmentally non-degrading, technically appropriate, economically viable and socially acceptable" (FAO 1989).

The concept of sustainability is very open to an interpretation that depends on the viewpoints and objectives of groups and individuals. Alley et al. (1999) state that resource sustainability is an elusive concept that cannot be defined in a precise manner and with universal applicability. While identifying a 'sustainable' practice, it is very important to state what it is that is being sustained. An important component of sustainability is determining indicators of performance relating to a set of goals. Because it is difficult to identify when a practice can be termed 'sustainable', indicators of sustainability are only useful when evaluating alternative operational or management practices. Indicators are an aggregation of information that indicates the change or defines the status of something. Indicators are most frequently based on quantifiable data, but may also be based on qualitative information depending on the purpose of the indicators (Gallopin 1997). Although well known and frequently used, there are several problems and difficulties related to the development of indicators; some of these problems are more technical, others are related to the use of indicators in the policy process. Several principles for creating legitimate indicators have been proposed, for instance by Hardi and Zidan (1997) and Meadows (1998).

### 2.1.2 Sustainable Agriculture

Sustainable agriculture is a subset of sustainable development. Many definitions of sustainable agriculture have been proposed, but one of the first to be adopted in the USA was published by the American Society of Agronomy (1989, p. 15):

> A sustainable agriculture is one that, over the long-term, enhances environmental quality and the resource base on which agriculture depends; provides for basic human food and fiber needs; is economically viable; and enhances the quality of life for farmers and society as a whole.

Another definition of sustainable agriculture reads as:

> Sustainable agriculture is both a philosophy and a system of farming. It is rooted in a set of values that reflects an awareness of both ecological and social realities and a commitment to respond appropriately to that awareness. It emphasises design and management procedures that work with natural processes to conserve all resources and minimise waste and environmental damage, while maintaining and improving farm profitability (MacRae et al. 1990).

The US Congress also defined sustainable agriculture in the 1990 Farm Bill (US Govt. 1990). Under that law, the term sustainable agriculture means an integrated system of plant and animal production practices having a site-specific application that over the long-term will:

- Satisfy human food and fiber needs.
- Enhance environmental quality and the natural resource base upon which the agricultural economy depends.
- Make the most efficient use of non-renewable resources and on-farm resources and integrate, where appropriate, natural biological cycles and controls.
- Sustain the economic viability of farm operations.
- Enhance the quality of life for farmers and society as a whole.

Thus, the US official definition has five parts, which emphasize productivity, environmental quality, efficient use of non-renewable resources, economic viability, and quality of life. Under this definition, a farm that emphasizes short-run profit but sacrifices environmental quality would not be sustainable in the long term. Conversely, pursuing environmental quality without ensuring viability of short-run returns would also be unsustainable. A farm that is very productive but uses large quantities of a non-renewable resource, such as fossil fuel or a non-rechargeable aquifer, to achieve and maintain that productivity would not be considered sustainable in the long term.

El Moujabber (Dr.) (undated) suggests that a system's perspective is essential to understanding sustainability. This approach gives us the tools to explore the interconnections between farming and other aspects of our environment. The author sees sustainable agriculture as both a philosophy and a system of farming that involves design and management procedures that work with natural processes to conserve all resources and minimize waste and environmental damage while maintaining or improving farm profitability. Furthermore, such systems aim to produce food that is nutritious, without being contaminated with products that might harm human health. Making the transition to a sustainable agriculture is a process, he says.

For farmers, the transition to a sustainable agriculture normally requires a series of small, realistic steps. Family economics and personal goals influence how fast and/or how far participants can go in the transition. The key to moving forward is the will to take the next step. The author also enlists five principles of agricultural sustainability:

- A sustainable agricultural system is based on the prudent use of renewable and/or recyclable resources. For instance, use of recyclable resources, such as groundwater, at rates greater than recharge depletes reserves and cannot be sustained.
- A sustainable agricultural system protects the integrity of natural systems so that natural resources are continually regenerated. For instance, sustainable agricultural systems should maintain or improve groundwater and surface water quality and regenerate healthy agricultural soils.
- A sustainable agricultural system improves the quality of life of individuals and communities.
- A sustainable agricultural system is economically profitable.
- A sustainable agricultural system is guided by a land ethic that considers the long-term good of all members of the land community.

Frank and Indu (2005) were of the view that the impacts of the rural sector on national development are far-reaching. Much of the natural resources concentrate in rural environments, and the mismanagement or degradation of those resources is closely linked to rural poverty. Since the term 'rural' traditionally has been, and to a large extent continues to be, popularly synonymous with agriculture, a sustainable rural development implies a sustainable agricultural development. The system theory, with application to agricultural systems, has been significantly progressed by the work of De Wit (Leffelaar 1999). According to De Wit, analyses of agricultural systems require interdisciplinary approaches. Like El Moujabber, he suggests that the cooperating disciplines may be placed into three main groups representing natural environment, social considerations, and financial aspects of the system (Fig. 2.2).

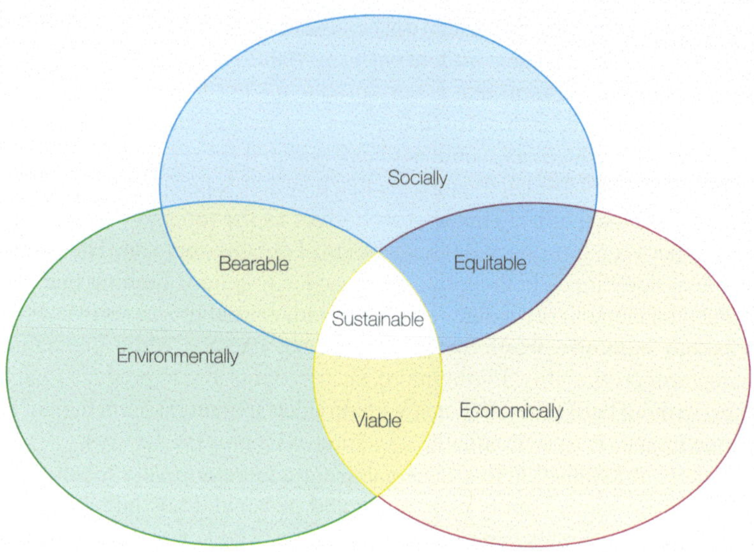


**Fig. 2.2** Three-tier model of a sustainable agricultural system

In the course of evaluating the progress of implementing Agenda 21, the UN 'Commission on Sustainable Development' defined sustainability as having not three but four dimensions (Spangenberg 2002) – adding institutions as the fourth dimension of sustainability. Zandstra (1993) reproduces a definition of sustainable agriculture given by the CGIAR in 1987 as:

> A sustainable agriculture is the successful management of resources for agriculture to satisfy changing human needs while maintaining or enhancing the quality of the environment and conserving natural resources.

He further writes that recently the Committee on Sustainability and Environment of CGIAR has adopted another definition for a sustainable agriculture, which states:

> A sustainable agriculture is one that over the long-term, enhances environmental quality and the resource base on which agriculture depends; provides for basic human food and fiber needs; is economically viable; and enhances the quality of life for farmers and society as a whole.

It is worthwhile to note that the American Society of Agronomy adopted the same definition in 1989. Zandstra further comments that sustainability is a comprehensive quality of a complex system; it is not a simple parameter to be measured directly. A possible way is to quantify the physical and biological processes occurring in a particular agro-ecosystem and quantify interactions between processes and components of the system. The next step would be to relate these quantities to environmental and social conditions and predict whether or not the system under consideration is heading in a sustainable direction (Zandstra 1993).

As sustainability cannot be measured directly, we need to identify measurable phenomena (indicators) that, when put together, suggest how sustainable our system

**Table 2.1** Goals of sustainability proposed by SCARM's expert group, 1993

| | |
|---|---|
| Overall goal | Maximize net profit over the long term |
| Economic goals | Optimize farm productivity |
| | Maintain contribution to the wider economy |
| Environmental goals | Hold and comply with resource consents |
| | Improve soil health |
| | Minimize adverse effects on water sources and receiving waters |
| | Minimize adverse effects on air |
| | Maintain or enhance biodiversity, habitats, and landscape |
| | Pursue effective waste management |
| | Minimize use of non-renewable energy resources |
| Social goals | Ensure acceptability of farming practices to the wider community |
| | Demonstrate good environmental management in the market place |

Source: Saeed (2012)

might be. In Australia, in 1992, the Standing Committee on Agriculture and Resource Management (SCARM) established an expert group to develop a set of indicators for sustainable agriculture to be used by decision makers at the regional and national scales. The group, in 1993, proposed that regional/national indicators could be divided into four main areas:

- Economic profitability;
- Land and water quality to sustain production;
- Managerial skills; and
- Off-site environmental impacts.

In the context of the sustainability of irrigated agriculture, the following 'sustainability goals' have been identified as being the key components of a sustainable agricultural system (Table 2.1). In cases, these goals can be conflicting. It is recognized that maximizing long-term net profit is the overall goal for farmers, but that this goal cannot continue to be met with the exclusion of the other identified economic, environmental, and social goals. The sustainability goals listed below have been identified as relevant to irrigated agriculture. Irrigation has many far-reaching effects on the environment that may not be apparent at first, so it is important that the effects on the whole system should be addressed from the beginning. It is therefore necessary to consider the relationship between these goals; considering any one goal in isolation will lead to an overall system that does not meet the overriding principle of sustainable management.

Pretty (1998) states that sustainability requires an integrated view of the world; it requires multidimensional indicators that show the links among a community's economy, environment, and society. He sets the following five goals for an agricultural system to be sustainable:

- A thorough integration of natural processes such as nutrient cycling and pest-predator relationships into agricultural production processes, so ensuring a profitable food production.

- A minimization of the use of those external and non-renewable inputs that have the potential to damage the environment or harm the health of farmers and consumers, and targeted use of the remaining inputs used with a view to minimize the costs of production.
- The full participation of farmers and other rural people in all processes of problem analysis and technology development, adaptation, and extension, leading to an increase in self-reliance among farmers and rural communities.
- A greater productive use of local knowledge and practices.
- The enhancement of wildlife and other public goods of the countryside.

### *2.1.3 Sustainable Use of Groundwater*

Agriculture is functionally linked to many other areas of environment. Among them, water is the most important resource for agriculture and that has played a pivotal role in the development of this sector; but it is also scarce and unevenly distributed both regionally and among certain specific marginalized populations. The current level of water use by agriculture may not be sustainable because of both its physical scarcity and competition for its use among other sectors, such as human consumption, sanitation, industry, and environment. As a result, therefore, many innovations to improve efficiency of water use are being tried. The benefits and costs of groundwater development change with time. In the early stages of economic and social development of an area, groundwater may play a vital role, particularly since it can allow smooth economic growth without the need for large previous investments. However, present circumstances may not be a good reference for the future. Therefore, cost–benefit analyses should be conducted within dynamic frameworks (Custodio 2002). The sustainable use of aquifers should be considered in a broad context of space, time, scientific status, available technology, and social development (Custodio et al. 2004). While groundwater development considerations have traditionally been associated with such terms as safe or perennial yield, the safe yield does not necessarily coincide with sustainability (Custodio 2002; Alley and Leake 2004). In many regions of the world, intensive groundwater use has helped to alleviate poverty; it is not known to be a cause of poverty. Conversely, it can be said that the most serious obstacles to sustainable groundwater development may be poverty and lack of democratic rules. Although some serious problems related to groundwater development (e.g. lack of water for the impoverished, general water quality impairment, or local disputes over the water supply) have been reported for extremely poor areas, these are overshadowed by more serious problems of another nature, including widespread illiteracy, authoritarian rule, social inequality, or corruption. A major threat to the sustainable use of aquifers is the deterioration of groundwater quality. However, deterioration in groundwater quality is only sometimes directly related to intensive groundwater use. Society's practices relating to the application of chemicals, in one form or another, have directly, and also indirectly, affected groundwater quality (Custodio et al. 2004). Frank and Indu

(2005) conclude that each groundwater system and development situation is unique and requires an analysis adjusted to the nature of the water issues faced, including the social, economic, and legal constraints.

The US geological department has set five priorities for sustainable groundwater management:

- Sustainable long-term yields from aquifers
- Effective use of the large volume of water stored in aquifers
- Preservation of groundwater quality
- Preservation of the aquatic environment by prudent abstraction of groundwater
- Integration of groundwater and surface-water into a comprehensive water and environmental management system

When the production of food and fiber degrades the natural resource base, the ability of future generations to produce and flourish decreases. The decline of ancient civilizations in Mesopotamia, the Mediterranean region, pre-Columbian Southwest US, and Central America is believed to have been strongly influenced by degradation of natural resources, principally from non-sustainable farming and forestry practices. Water is the principal resource that has helped agriculture and society to prosper, and it has equally been a major limiting factor when mismanaged.

## 2.2 Literature Review

For better understanding of the conclusions that various scholars have drawn from their studies, the literature review is arranged in two segments.

### *2.2.1 Agriculture-Water Nexus: International Literature*

Agriculture and water are functional twins. Various interrelated aspects of agriculture and water, as irrigation resource, have been studied by researchers across the world (e.g. Sadler and Cox 1987; Evans and Cant 1981; Rosenzweig and Hillel 1995; Wolf 1995; Postel 2003; Scanlon et al. 2005; Pennell 2006 etc.). Sadler and Cox (1987) suggest that water management is an important component of socio-political processes. Water resource management programs depend on natural conditions, as well as the water needs determined by prevailing socio-economic conditions among societies. Under favorable water resource conditions, or low levels of socio-economic development, water management may not be a significant activity, but under difficult conditions in that respect, water management generally becomes an important agenda point for governmental policies and practical strategies. They further say that the traditional approach of developing additional supplies of water from a single source is unsustainable. The need is for 'demand control' by legislative and educative means, and by the use of economic incentives. Adoption of water pricing mechanisms and price structures designed to limit use has significant potential to encourage water recycling and other water-saving practices. For a problem of salinity management in Australia, they have found a significant role for socio-political factors in water resource management. They have concluded that as water demands increase with an expanding population and water-use activities, water management generally will increase. For a higher degree of effectiveness, water management is likely to require modification of the traditional view that it is primarily a technical activity simply involving construction of engineering works to modify hydrologic systems. Rather, water problems must be viewed in their broad social contexts. They have also found that, without local support and approval, water management activities are not feasible, and that the source of a water problem is deeply associated with the problems of land management.

Evans and Cant (1981) have studied the economic and social impacts of irrigation. In a study in New Zealand, they found that capital value of land and settlement density increased with an increase in irrigated agriculture, while farm size decreased with irrigation advancement. Further, through their study they came to view irrigation as necessary for agricultural intensification and minimisation of the risks associated with climatic uncertainties. They also quote from Stewart (1963) his finding that the financial returns achieved on irrigated farms were no better than those on non-irrigated farms where irrigation had a higher cost in areas where rainfall is sufficient for crop productions. Wolf Peter (1995) argues that sustainability of irrigation systems should always be viewed in conjunction with the institutions that operate the systems. It is impossible to guarantee the sustainability of any part of the irrigation structure if no institution exists to ensure proper operation, maintenance, and modifications. It indicates certain socio-economic and financial risks, like ineffective institutions, economic competition (cost-benefit gap), manpower inadequacy, etc. and concludes by suggesting that the performance efficiency of irrigation systems be continuously monitored against risks. The author quotes from Abernethy (undated), who lists sustainability threats as when:

(a) The respective system no longer offers any economic advantage.
(b) The respective system operates extremely well internally but externally has a disadvantageous effect on other interests.
(c) The people involved are no longer willing to bother about major aspects of the system.
(d) The systems in general come under severe pressure.

The term over-exploitation has been frequently used in the last three decades. Nevertheless, most authors agree in considering that the concept of aquifer over-exploitation is poorly defined and resists a useful and practical definition (Sophocleous 1997, 2000). A number of terms related to over-exploitation can be found in water resource-related literature. Some examples are safe yield, sustained yield, perennial yield, overdraft, groundwater mining, exploitation of fossil groundwater, optimal yield and others (Adams and MacDonald 1995). In general, these terms have in common the idea of avoiding 'undesirable effects' as a result of groundwater development. However, this 'undesirability' depends mainly on the social perception of the issue. This social perception is more related to the legal, cultural, and economic background of a region than to its hydro-geological facts. In an article, Hassan and Garg (2003) mention low efficiency and the financially non-sustainability of groundwater irrigation. The article presents an estimate showing that, with a 10 % increase in the present level of water-use efficiency in irrigation projects, an additional 14 mha area can be brought under irrigation from the existing irrigation capacities. It quotes a policy of water pricing that should be such as to convey its scarcity value to the users and motivate them in favor of efficient water uses. Nikhel et al. (2003) provide certain techniques for harvesting of rainwater and advocate that utilization of irrigation water in efficient and economical ways will conserve lot of water to utilize for irrigating larger areas. They recommend adoption of advanced means of irrigation such as drips, lines, sprinklers, etc., and the adoption of a scientific approach for successful water resource management. Ali (2003) suggests two ways for achieving water conservation:

1. Efficient use of available water.
2. Recycling/reuse of the water.

According to him, these two objectives can be achieved via technology and public education. In the developing world, agriculture is the largest water consumer. This is because of:

- Inefficient irrigation systems.
- Lack of education and training of farmers.

The main hurdles in changing the irrigation system for better are financial constraints, unawareness, and poor legislative control. Priyan (2003) says that there is a strong link between water scarcity and the way water is used for irrigation. It suggests that technical analysis of various aspects, like water-balance study, water accounting, etc. are required to augment the supply of water. But technical interventions alone are not fruitful unless demand is managed. Sustainable and environmentally friendly irrigation practices help to achieve growth in agricultural productivity. The author suggests several measures for ensuring water security. The measures include demand-based management, socio-economic analysis, technological and institutional interventions, conjunctive use of surface- and groundwater, adoption of efficient irrigation practices, decentralization of water management activities, training of farmers in water conservation, volumetric pricing of water, etc. Ponce (2006, 2007) suggests that the sustainability of groundwater utilization must be assessed from an interdisciplinary perspective, where climatology, hydrology, ecology, and geomorphology play an important role. He has discussed some very striking features of groundwater. For instance, groundwater is a common pool resource, the unregulated use of which leads to the 'tragedy of the commons', with the eventual depletion of the resource and ruin to all. Depletion in one aquifer has the tendency to draw from neighboring connected aquifers and under such conditions overpumping may have a large area of influence. The impacts of groundwater depletion include drying up of wells, increased cost of pumping and well infrastructure, land subsidence, salt-water intrusion, and changes in surface albedo, and related climate change. He also suggests that sustainability in water quantity must imply sustainability in water quality. In rural areas, inappropriate land use and other human activities can lead to aquifer contamination. Prasad Narasimha and James (2003) provide a key idea that the impact of any irrigation project on groundwater must be verified only after the rainy season and during the irrigation period.

In a study on sustainable agricultural policies, Wilson and Tyrchniewicz (1995) developed an analytical set of principles for a sustainable agriculture (Table 2.2). They tested the merits of this analytical framework by using it to assess the compatibility of four policies with sustainable agriculture. In the framework, the primary instruments included subsidy, supply management, contracts, and financial incentives. Subsidy as applied under the 'Western Grain Transportation Act' was found to be inconsistent with sustainable agriculture. Supply management as exercised under the Farm Products Marketing Agencies Act with respect to eggs was found neutral to or consistent with some of the principles of sustainability while being inconsistent with others. The contract instrument as utilized under the Prairie Farm Rehabilitation Act with respect to the Permanent Cover Program was found to be consistent with sustainable agriculture. The financial incentive instrument, as used under the North American Waterfowl Management Plan, was found to be consistent with sustainable agriculture. Regarding the interconnections between the ingredients of sustainability, the authors conclude that:

- Environmental damage is of importance not only because of its economic effects but because it lowers productivity; soil degradation being one example.
- A greater priority must be given to the environment if economic policies are to be sustainable; such as in agricultural policy.
- Attention is required to ensure that economic growth is also sustainable growth.
- While raising income continues to be a major goal of policy, the need remains to be aware of the potential trade-offs between increases in income and environmental deterioration, with poverty understood to be a root cause of environmental degradation.

The survey conducted by FAO (1995) estimates that, of the total annual freshwater withdrawals, the world average for agricultural use is 70 %. On a regional basis, the proportion of water abstracted by the agricultural sector is highest in less developed regions. In Asia, for instance, the share of agriculture in the total annual freshwater withdrawals is 81 %. This means that agricultural sectors consume larger quantities of water in less developed countries than in developed countries. This situation is attributed to less efficient irrigation systems in the less developed countries. Pennell (2006) provides a review of the problems of forming a sound policy for management of dry land salinity in Australia. His article brings together considerations of hydrology, farmers' perceptions and preferences, and farm-level economics of salinity management practices, etc. Scanlon et al. (2005) studied the changes in the groundwater recharge potential and the flushing of salts to the underlying aquifers affected by changes in natural land cover due to

**Table 2.2** Principles of sustainable agriculture, by Wilson and Tyrchniewicz (1995)

| | |
|---|---|
| 1. Management | Maintain the integrity of ecosystems |
| | Enhance the (quantity and quality) flow of services from the resource base for present and future generations |
| | Provide for integrated (shared) resource management |
| 2. Conservation | Efficient use (consumption) of all resources, both renewable and non-renewable |
| | Maintain biological diversity |
| | Provide habitat for wildlife and plants both on land and water |
| | Optimum use of land for sustainability |
| 3. Rehabilitation | Restore the productivity of a degraded resource |
| | Apply waste management principles (reduce, reuse, and recover) |
| | Promote complementary production systems |
| | Promote closed production systems where appropriate |
| | Replace degrading processes with others that are beneficial |
| | Revitalize the resource |
| 4. Market viability | Reduce trade barriers |
| | Use resources in an economically efficient manner |
| | Assure a sustainable income |
| | Promote sustainable human economic activity |
| | Sensitive to supply and demand in the market place |
| | Unbiased as to commodities and mode of transport |
| | Enhance value-added activity |
| 5. Internalization of cost | Promote full environmental costing |
| | Include all costs associated with economic activity |
| | Contingent valuation where costs cannot be internalized |
| | Use of natural system economic accounting (inclusion of resources and externalities in system of national accounts) |
| | Assess beneficiaries of externalities |
| | Appropriate costs |
| 6. Scientific and technological innovations (R & D) | Enhance air, water, and land management |
| | Ameliorate waste management |
| | Increase productivity |
| | Reduce consumption of non-renewable resources |
| | Promote technology transfer |
| | Advance biotechnology |
| | Promote technologies that utilize yet preserve native ecosystems |
| | Promote technologies to further environmental quality, including human health and economic growth |
| | Develop industries benign to environment |
| 7. Trade policy | Maintain or enhance resource base of different trading regions |
| | Apply true comparative advantage |
| | Promote international market responsiveness |
| | Increase value-added exports |
| | Consistent with trade agreements |
| | Support trade agreements that recognize externalities |

(continued)

Table 2.2 (continued)

| | |
|---|---|
| 8. Societal considerations | Promote gender equity |
| | Enhance human health and education |
| | Preserve aesthetic values |
| | Water quality and quantity available for alternative uses |
| | Alternative options for employment (adjustment programs) |
| | Maintain and/or enhance food quality, safety, and quantity |
| | Societal neutrality (does not privilege one group over another) |
| | Protect agriculture from urbanization |
| | Increase productive capacity of the poor |
| | Promote fairness and equity in resource allocation for commercial and recreational purposes |
| | Provide an acceptable quality of life and livelihood |
| | Sensitive to goals of local people and communities |
| | Respect human rights |
| 9. Global responsibility | Recognize interdependence among nations |
| | Promote intra- and intergenerational equity |
| | Encourage food health and safety |
| | Assist in emergency food aid |
| | Promote technology transfer – research and development |
| | Promote fairness and equity in income distribution and trade |

Source: Saeed 2012 (Derived from Wilson and Tyrchniewicz 1995)

agricultural expansion in the Amargosa Desert in the USA. Moreover, studies have investigated irrigated cropland in the high plains in the USA (Kettle et al. 2007) and the Rize area in northeast Turkey (Reis 2008). These studies used satellite images to detect changes in land cover and agricultural land use. The study of Reis, for example, concludes that the increase in agricultural land use mostly result in deforestation, which means some forest areas were removed and converted to farming in the region. Rosenzweig and Hillel (1995) theoretically enlisted critical environmental factors, including climate and water availability, which can bring about drastic changes in agricultural development if they are changed from any standard norm. Sen (2004) undertook a very interesting study in Portugal, combining information from remote sensing and GIS and developing a water-balance model of the Roxo Dam catchment via the Thornthwaite-Mathew method, which is based on recent climatological data, to assess available water for irrigation. Postel (2003) emphasized strengthening of irrigation-related institutions in order to ensure a judicious water use in agriculture. Marsh (undated), in a study in India, found that the highly subsidized flat rate for electricity encouraged the use of water in a wasteful manner in tubewell irrigated agriculture.

Rosegrant and Cai (2002) assessed the impacts of water supply limitations on growth of food production in the future. Their work developed a global modeling framework, named 'IMPACT-WATER' that combined an extension of the International Model for Policy Analysis of Agricultural Commodities and Trade (IMPACT) with the Water Simulation Model (WSM). The IMPACT-WATER model estimates domestic and industrial water demands as a function of population and income. Water demand in agriculture is projected based on irrigation and livestock growth, climate, and base-level irrigation water-use efficiency. Mosher (undated) recommends three basic ways in which food production can and will be increased within the next generation. The first of those ways is taking the steps that will lead to a takeoff in yield per unit of land on the presently cultivated area in the less developed countries. The second way is the substantial expansion of cultivated area. He admits that this second step will be more difficult and more expensive because it will require extensive groundwater and hydrologic surveys, and construction of irrigation infrastructure to allow double and triple cropping compared with that which is present in much more lands of South and Southeast Asia. However, the present indicators suggest that it may be possible to double the present magnitude of globally cultivated area by bringing more of the available cultivable lands under plough, primarily in the continents of Africa and South America. The third possibility that Mosher sees in future is the expansion of cultivation onto seas, and the chemical synthesis of foods.

Custodio et al. (2004) demonstrated that one of the major threats to the sustainable use of aquifers is the deterioration of groundwater quality. However, in their view, groundwater quality deterioration is rarely directly related to intensive groundwater use. Rather, in most cases, the society's practices relating to the application of agro-chemicals in one form or another have directly and/or indirectly affected groundwater quality. Intensive groundwater use requires adequate management as a necessary step to ensure sustainability and long-term beneficial use of groundwater. They go further by saying that management plans should encompass not only the technical aspects of sustainability but also its economic, social, cultural, and environmental concerns. The authors suggest that groundwater management should be applied at the 'local' level (i.e. lower administrative and territorial level) rather than at a higher level. Public participation is essential for the successful implementation of a groundwater management plan. When public participation is limited to certain interest groups, or to certain stages of the management process, this type of public participation is not truly effective. They further suggest that there are some limitations to a purely regulatory approach to groundwater management. For instance:

1. A rigid licensing system, where groundwater abstraction permits are granted for specific uses in specific locations, does not provide the necessary flexibility to address situations of stress or social change adequately. A more flexible regulatory framework would allow for temporary or permanent transfers of water rights between users, so that efficiency and equity criteria could be met.
2. Effective regulatory approaches require the existence of adequate enforcement tools and institutions to ensure compliance.
3. A groundwater management system, which is based primarily on regulatory means, also requires the existence of a complete and continuously updated inventory of water licenses or rights, and reliable and generally accepted information on available resources and existing rights. When adequate information is not available, participatory management tools are required to guarantee social acceptance and the effective implementation of management programs.

4. Existing water legislation may no longer serve current social demands for water resources, particularly in areas where water supplies are scarce or intensive groundwater development already exists. Therefore, a comprehensive water management plan is particularly necessary to address these conditions.

Custodio et al. are of the view that dominantly punitive regulations, which are common in many situations, may not be enough. They may hinder, rather than help, achieve the goals of a groundwater management plan. Hence, special attention should be paid to educating farm managers, policy makers, and other stakeholders in the special characteristics of groundwater. The attentive measures in this regard specially include providing an understanding of the relatively slow response of groundwater systems to physical stresses. It is also important that these groups should understand the even slower response that occurs when remedial measures are implemented to mitigate groundwater contamination and the generally high cost of these measures (Custodio et al. 2004).

### 2.2.2 Agriculture-Water Nexus: National Literature

Balochistan is considered the 'fruit basket' of Pakistan. During the last two decades or more, a number of Government-sponsored documents have emerged addressing the water resource issues of the province, but scientifically designed and executed analytical research articles/theses are very rare in this field. Khan and Nawaz (1990) undertook a UN-sponsored study project on the land use and small-scale irrigation systems in the Kech Valley of Balochistan. The project concluded that the Government had no active role in the management of these systems. Rather, such community-based small-scale irrigation schemes are governed by traditional customs and mutual social cooperation. Another article, by Morton and Hoeflaken (1995), explores performance of the flood irrigation schemes in Balochistan. It found that Government involvement in such schemes has not been one of participation by the supposed beneficiaries. Thus, many schemes are a failure either due to gross technical flaws, or a combination of technical and social reasons. The social factors of failure that they have highlighted include poor management, irrigation-based inter-village and intra-village conflicts, misleading influences of tribal leaders, ethnicity, etc. The technical defective factors include inadequate designs and locations of the flood irrigation schemes. Water user's associations are not found anywhere in Balochistan. In Pishin District only, the Government imposes some tax for cost recovery of the irrigation schemes (Morten and Hoeflaken 1995).

The study of Achakzai and Toor (1989) explains the technology of karez systems, and the rules of water allocation and distribution within them. It concludes that local institutions are gradually losing control over equitable water management because of the increasing scarcity of the resource and social changes due to which co-operational spirit among stakeholders is evaporating. One other study has found that tubewell technology is emerging as the dominant source of irrigation at the expense of other traditional sources (e.g. karez), mainly because of the higher levels of social and economic inequality among the water users (Syed 1991). Although legislation that prohibits installation of a new tubewell nearer than a certain criterion distance to another tubewell or karez exists on paper, it is commonly ignored due to institutional deficiencies (Syed 1991). Overstepping of such restrictions sometimes leads to serious social conflict (Gills and Baig 1992). The literature also revealed that the community-based irrigation systems (karezes, mountain springs, etc.) are being gradually substituted by individually managed systems (dugwells and tubewells), primarily due to disruption in the traditional social norms of paying sincerity to collective action calls (Syed 1991).

Different studies establish that over the last three decades, groundwater resources in most of Balochistan Province have dwindled remarkably. For instance, according to Saeed (2006), water depletion in the Nari and Pishin-Lora river basins of Balochistan was noted since the 1970s. However, during the last 10–15 years, the groundwater in these two basins depicts a perfect case of over-abstraction and there is no hope of recharge for them in years to come. His study further concludes that, in many areas of Balochistan, agriculture, as an occupation and business, is no longer a profitable enterprise, as it was 10–15 years ago. He has highlighted a number of reasons for that:

- Natural calamities, like the continuous drought during 1997–2004.
- Institutional weaknesses.
- The high cost of agricultural inputs like fertilizer, pesticides, and transportation.
- Explosive increase in number of tubewells, contributing to the over-exploitation of groundwater.
- Growing of high-delta crops, like apples in Pishin Valley and onions in Mastung and Kalat regions.
- Drying out of karezes and springs, which were considered sustainable systems.
- Defective irrigation systems, like flooding the entire orchards with water instead of using the HIESs (e.g., trickle, bubbler systems).
- The heavily subsidized flat rates of electricity for irrigation tubewells, which tempts farmers to install more tubewells and keep them operating without an off switch.

A report on the Integrated Water Resources Management (IWRM) Policy for Balochistan (GoB 2004) lists five major reasons for groundwater mining:

- Continuous increase in water demand.
- Lack of incentives for efficient water-use due to heavily subsidized electricity tariff.
- Unsustainable magnitude of groundwater abstraction.
- Neglect of natural recharge systems, and
- Ineffective and expensive techniques for inducing recharge to the groundwater in the name of DADs.

Nawaz (2004) points out the following three reasons for the phenomenal increase in private tubewells, which started in the 1970s:

- The flat rate policy in electricity tariff for agricultural tubewells.
- Advent of modern deep drilling technology.
- Government policy of greening the province, i.e. financially supporting groundwater development schemes.

His study claims that the desertification process has started in the Uplands of Balochistan due to decline of the watertable, and that the over-use of groundwater may result in its poor quality. The author asserts that although Balochistan has considerable floodwaters, they are not properly managed and utilized for agriculture. There are enough promising experiences of community-based floodwater management systems in the province, which are not only low cost but also environmentally friendly. The revival/rehabilitation of those systems would be a good option in resolving the growing water crisis. Van Steenbergen and Oliemans (undated) suggest that until the 1960s, groundwater supplies were considered limitless in the province. The Government promoted well development by providing electricity at cheaper rates. Also, to make it easy to collect dues, the Government applied a system of flat rates in tariff to most electrified tubewells. This move encouraged over-pumping, as the electricity charges bore no relation to consumption. The authors also found that drying out of karezes had both hydrological and social components. When increased withdrawal from the new dug wells accelerated the lowering of the groundwater tables, karezes became less viable, and the first people to release their share in the communal systems were often the larger farmers who had the resources to develop a private well(s). The heavy burden of maintaining the drying karezes then fell increasingly upon the smaller farmers. The final outcome was often the collapse of the traditional system and a polarization of the access to groundwater. The authors have also questioned the effectiveness of the DADs, because the reservoirs tend to quickly silt up, thus impeding recharge and reducing storage capacity. In terms of roundwater laws, the authors conclude that in most of the valleys, a 'help yourself' attitude prevailed; the ordinance and its provisions were ignored in all cases Van Steenbergen and Oliemans (undated). Syed (1991) has discussed various aspects of the karezes, which are Balochistan's typical irrigation systems. The author has also spoken about some groundwater-use laws in Balochistan, which provides restrictions for installation of tubewells at a certain distance from a karez, but he categorically states that such rules are commonly ignored due to the lack of enforcement capacities in institutions, lack of political will, and corruption.

As mentioned above, we found no elaborate research work to have scientifically studied the progress of agricultural development in a scenario of groundwater depletion in the peculiar physical, socio-political, and policy environment of Balochistan. The proposed study is primarily intended to fill this gap. In addition, it will build on previous work addressing aqua-agro relationships worldwide by investigating an area that is undergoing rapid changes in both irrigation resource availability and agricultural land use.

## 2.3 Summary

This chapter conceptualized the problem by its various angles, thus providing a basis for research design. It linked the goal of sustainable rural development with systems theory, which is used to guide decisions on resource management, regional development, and technology development, etc. It discussed that sustainability integrates three main goals – environmental health, economic profitability, social equity, and also quoted scholars having added a fourth goal – institutional strength. Here it was established that sustainability of a system cannot be measured directly; rather some measureable phenomena can indicate it. It proposed that sustainable agriculture enhances the resource base on which it depends (e.g. water), provides basic human food and fiber needs, and is economically profitable. The chapter revealed that benefits and costs of groundwater exploitation change with time; in the early stage of an area's development, groundwater may play an essential role as it can allow smooth economic growth without large previous investments. It described that a major threat to aquifer sustainability is groundwater quality deterioration. It was also highlighted that each groundwater system and development system is unique, thus requiring an analysis specially adjusted for it, including social, economic, and legal constraints. The literature suggested that the main hurdles in improving an irrigation system are financial constraints, unawareness, and poor legislative control. For Balochistan, studies have discussed that lack of political will, institutional weakness, and corruption are the main causes behind agricultural and water resource deterioration. The literature survey further revealed that little research has been conducted in the peculiar physical, social, and policy environment of Pishin Valley and the rest of Balochistan. The literature available comprises mostly bureaucratic documentations, reports of donor-funded projects, or simply descriptive journal articles.

## References


Achakzai GN, Toor AS (1989) Patterns of management in the Karez system of Balochistan. Pakistan Council of Research in Water Resources, Islamabad, pp 70–78

Adams B, Macdonald A (1995) Over-exploited aquifers. Final report, British Geological Survey, technical report WC/95/3, p 53

Alcamo J, Kreileman GJJ, Krol MS et al (1994a) Modeling the global society–biosphere–climate system: part 1: model description and testing. National Institute of Public Health and Environmental Protection, Bilthoven, Accessed at: http://sedac.ciesin.org/mva/KLUWER1994/AlcamoJosephed.1994.IMAGE2.0.html. Accessed 23 Sep 2012

Alcamo J, Van Den Born GJ, Bouwman AF et al (1994b) Modeling the global society–biosphere–climate system: part 2: computed scenarios. National Institute of Public Health and Environmental Protection, Bilthoven, Accessed at: http://sedac.ciesin.org/mva/KLUWER1994/image-2.0-kluwer-ch2.html. Accessed 23 Sep 2012

Ali A (2003) Water conservation in New Era. In: Water and wastewater perspectives of developing countries, Conference proceedings. Anamaya Publishers, New Delhi

Alley WM, Leake SA (2004) The journey from safe-yield to sustainability. Groundw J 42(1). National Groundwater Association, USA. Accessed at: http://info.ngwa.org/gwol/pdf/040678070.pdf. Accessed 24 Jul 2013

Alley WM, Reilly TE, Frank OL (1999) Sustainability of groundwater resources. U.S. Geological Survey, Circular 1186, Denver, CO

American Society of Agronomy (1989) Decision reached on sustainable agriculture. Agronomy News. Madison, Wisconsin, January 1989, p 15

Costanza R, Patten BC (1995) Defining and predicting sustainability. J Ecol Econ 15(3):193–196, Elsevier. Available at: http://www.sciencedirect.com/science/article/pii/0921800995000488. 25/7/2013

Custodio E (2002) Aquifer overexploitation: what does it mean? Geological survey of Spain. Hydrogeol J 10(2). Madrid. Accessed at: http://www.yemenwater.org/wp-content/uploads/2013/03/Custodio-Emilio-2002.pdf. Accessed 15 Dec 2010

Custodio E, Kretsinger V, Llamas MR (2004) Intensive development of groundwater: concept, facts and suggestions. Water Policy J 7. Accessed at: http://www.rac.es/ficheros/doc/00248.pdf. Accessed 15 Dec 2010

El Moujabber M (undated) Sustainable agriculture. Accessed at: http://www.sowamed.ird.fr/resource/RES310_WP6_IntroductionSustainableAgriculture.pdf. Accessed 15 May 2009

Evans M, Cant G (1981) The effect of irrigation on farm production and rural settlement in Mid-Canturbury: a comparison of the irrigated and dryland farming zones in the Lyndurst-Pendarves area, 1945–1946. J N Z Geogr 37(2):58–66, New Zealand

FAO (1989) Arid zone forestry. Conservation guide no. 20. Rome

FAO (1995) Survey sheet, land and water division (AQUASTAT). Accessed at: http://www.fao.org/docrep/015/i2490e/i2490e04b.pdf. Accessed 26 Jul 2013

Frank B, Indu JA (2005) Harvesting opportunities: rural development in the 21st century. Concept Papers, Fourth Regional Thematic Forum, World Bank

Gallopin GC (1997) Indicators and their use: information for decision-making. In: Billhartz SB, Matravers R (eds) A report on the project on indicators of sustainable development. Wiley, Chichester, pp 13–27

Gills VH, Baig MS (1992) Environmental profile of Balochistan. Land Resources and Urban Science Department/ Ecology Unit, Soil Survey of Pakistan, Netherlands/Lahore

GoB (2004) Integrated water resources management policy, Balochistan, Pakistan: final draft report, Component # 3, Balochistan Resource Management Programme

Hardi P, Zidan T (1997) Assessing sustainable development: principles in practice. International Institute for Sustainable Development, Winnipeg

Hassan Q, Garg NK (2003) Management of water resources: an integrated perspective. Printed in Water and Wastewater Perspectives of Developing Countries, conference proceedings, Devi R, Ahsan N (eds) . Anamaya Publishers, New Delhi

Kettle N, Harrington LMB, Harrington JA (2007) Groundwater depletion and agricultural landuse change in the high plains: a case study from Wichita county, Kansas. Prof Geogr J 59(2):221–235, Accessed at: http://krex.k-state.edu/dspace/bitstream/handle/2097/4947/KettlePG2007.pdf?sequence=1. Accessed 18 Jun 2013

Khan MFK, Nawaz M (1990) Irrigation and landuse in the Kech Valley, Makran. Pakistan Council of Research in Water Resources, Islamabad, unpublished

Leffelaar PA (1999) On systems analysis and simulation of ecological processes; with examples in CSMP, FST and FORTRAN, vol 4, Current issues in production ecology. Kluwer Academic Publishers, New York, p 318

MacRae RJ, Hill SB, Hennings J, Bentley AJ (1990) Policies, programs and regulations to support the transition to sustainable agriculture in Canada. Cambridge University Press, Cambridge. Am J Altern Agri (now known as Renewable Agriculture and Food System) 5(2)

Malafant KWJ, Fordham DP (1997) Integration frameworks in agricultural and resource planning and management. In: Munro RK, Leslie LM (eds) Climate prediction for agricultural and resource management. Australian Academy of Science, conference, Canberra, Bureau of Resource Sciences, Canberra, Australia. Accessed at: http://www.complexia.com.au/Documents/AAS_Paper/AAS_Pape.html. Accessed 20 July 2013

Marsh D (undated) Estimating the price of water in Pondicherry, India: a framework for policy analysis. Department of Economics, University of Waikato, Hamilton, New Zealand. Accessed at: http://www.mssanz.org.au/MODSIM99/Vol%202/Marsh.pdf. Accessed 28 Jun 2009

Meadows D (1998) Indicators and information systems for sustainable development. The Sustainability Institute, Hartland, Accessed at: http://www.iisd.org/pdf/s_ind_2.pdf. Accessed 12 July 2013

Morton J, Hoeflaken VH (1995) Some findings from a survey of flood irrigation schemes in Balochistan, Pakistan. Water Resour J. Economic and Social Commission for Asia and the Pacific, UNO

Mosher AT (undated) World food problems and prospects. http://ageconsearch.umn.edu/bitstream/17572/1/ar670021.pdf. Accessed 13 July 2008

Nawaz K (2004) Community participation and water management in Balochistan, Pakistan. http://archive.unu.edu/env/workshops/Aleppo/06%20-%20Nawaz.doc. Accessed 16 May 2008

Nikhel K, Sundararjan M, Misra DD (2003) Integrated water resource management for the Jhakrand State: a conceptual model, Printed in Water and wastewater perspectives of developing countries, Conference Proceedings, Devi R, Ahsan N (eds). Anamaya Publishers, New Delhi

Pennell JD (2006) Dryland salinity: economic, scientific, social and, policy dimensions. Aust J Agric Resour Econ 4(45):517–546

Ponce VM (2006) Groundwater utilisation and sustainability. Accessed at: http://groundwater.sdsu.edu/. Accessed 24 Jul 2013

Ponce VM (2007) Sustainable yield of groundwater. Accessed at: http://ponce.sdsu.edu/groundwater_sustainable_yield.html. Accessed 24 Jul 2013

Postel SL (2003) Securing water for people, crops, and ecosystems: new mindset and new priorities. Nat Res Forum 27(2). Blackwell Publishing, Oxford. Accessed at: http://www.aseanenvironment.info/Abstract/41015659.pdf. Accessed 8 Jun 2010

Prasad Narasimha NB, James EJ (2003) Impact of an irrigation project on groundwater regime within a command area – a case study. Printed in Water and Wastewater Perspectives of Developing Countries, conference proceedings, Devi R, Ahsan N (eds). Anamaya Publishers, New Delhi

Pretty JN (1998) Toward more conducive policies for sustainable agriculture. Published in Agriculture and the Environment: Perspectives on Sustainable Rural Development, A World Bank symposium, p 37

Priyan K (2003) Emerging issues in water management and strategies for water security in Gujrat, India. Printed in Water and Wastewater Perspectives of Developing Countries, Conference Proceedings, Devi R, Ahsan N (eds). Anamaya Publishers, New Delhi

Reis S (2008) Analyzing land use/land cover changes using remote sensing and GIS in Rize, north-east Turkey. Sensors—Open Access Journal, published online by MDPI AG, Switzerland. Available at: http://www.mdpi.com/1424-8220/8/10/6188/pdf

Rosegrant MW, Cai X (2002) Water constraints and environmental impacts of agricultural growth. Am J Agric Econ 3(84):832–838, Oxford University Press, USA

Rosenzweig C, Hillel D (1995) Potential impacts of climate change on agriculture and world food supply. Consequences 1(2):23–32

Sadler BS, Cox WE (1987) Water resources management: the socio-political context. Water Res J Econ Soc Comm Asia Pac. UNO

Saeed M (Dr.) (2006) Promising crops and water efficient cropping patterns for irrigated farming systems of Balochistan, Pakistan. Summary report, TA-4560 (PAK), Project for Supporting Implementation of IWRM Policy in Balochistan. Govt. of Balochistan, ADB, and Royal Netherlands Govt

Saeed A (2012) Inter-effects of tubewell irrigated agriculture and the aquifer potential in Balochistan: a case study of Pishin Valley, 1981–2008. Ph.D. thesis, University of the Punjab, Lahore, Pakistan

Scanlon BR, Reedy RC, Stonestrom DA, Prudic DE, Dennehy KF (2005) Impact of land use and land cover change on groundwater recharge and quality in the southwestern US. Glob Chang Biol 11:1577–1593, Blackwell Publishing Ltd

Schoute JFT, Finke PA, Veeneklass FR, Wolfert HP (1995) Scenario studies for the rural environment. Kluwer Academic Publishers, Dordrecht

Sen PK (2004) Diagnosing irrigation water resources with multi-sensor remote sensing and GIS techniques: a case study of the Roxo Dame irrigation system, Portugal. M.Sc. Thesis, ITC, Enschede, The Netherlands

Sophocleous MA (1997) Managing water resources systems: why "safe yield" is not sustainable. J Ground Water 35(4):561, USA. Editorial available at: http://ponce.sdsu.edu/sophocleous_gw1997.pdf, 24/7/2013

Sophocleous MA (2000) From safe yield to sustainable development of water resources – the Kansas experience. J Hydrol 235(1–2):27–43, Elsevier, USA. Available at: http://people.ucalgary.ca/~hayashi/geog515/reading/Sophocleous2000.pdf, 24/7/2013

Spangenberg JH (2002) Environmental space and the prism of sustainability: frameworks for indicators measuring sustainable development. J Ecol Indic 2(3):295–309, Elsevier

Swart RJ, Raskin P, Robinson J (2004) The problem of the future: sustainability science and scenario analysis. J Glob Environ Chang Part A 14(2):137–146, Elsevier

Syed MA (1991) The diffusion of tube-well technology and karez abandonment in Balochistan. Selected papers, Agricultural Economy of Balochistan, Quetta Printing Press, Quetta

US Government (1990) Food, agriculture, conservation, and trade act of 1990, Public law 101-624. Title XVI, Subtitle A, Section 1603. Washington, DC

Van Ittersum MK, Rabbinge R, Van Latesteijn HC (1998) Exploratory landuse studies and their role in strategic policy making. J Agric Syst 58(3):309–330, Elsevier

Van Steenbergen F, Oliemans W (undated) Groundwater resource management in Pakistan. Accessed at: http://www.groundwatermanagement.org/documents/14locgrounwregulationpraxis.pdf. Accessed 12 Mar 2008

WCED (1987) Our common future: the Brundtland Report, World Commission on Environment and Development, UN, Oxford University Press

Wilson AJ, Tyrchniewicz A (1995) Agriculture and sustainable development: policy analysis on the great plains. International Institute for Sustainable Development, Winnipeg, Accessed at: http://www.iisd.org/pdf/agri_sd.pdf. Accessed 24 Nov 2009

Wolf P (1995) The problem of the sustainability of irrigation systems. Institute for Scientific Co-operation, Tubengen, J App Geogr Dev 45/46

Zandstra (1993) Preserving the options: International research for sustainable agriculture. In: Agriculture and environmental challenges. Proceedings of the 13th agriculture sector symposium, Srivastava and Alderman (eds). World Bank, Washington, DC, pp 101–107

# Chapter 3
# Environment of the Study Area

**Abstract** An agricultural system can be viewed as a unit ecosystem, the study of which requires a thorough understanding of the physical, social, political, and economic background. This chapter, therefore, contains baseline information about the study area regarding the existing status of some key environmental elements, including topography, geology, climate, soil, drainage, demography, tenure culture, etc., which can all make a difference. To start with, the chapter introduces the area in terms of location, size, and administrative affiliation. It details the relief and lithology (as they are crucial in aquifer discharge and recharge potential), and discusses soil and climatic characteristics (which have critical bearings upon the agricultural system). Further, the farming society has been introduced for its demographic features, socio-economic status, and political organization.

**Keywords** Balochistan • Pishin Valley • Location • Climate • Population

## 3.1 Location, Size, and Administration

The Pishin Valley is bounded by latitudes 30° 13′ and 30° 47′ N and longitudes 66° 27′ and 67° 13′ E. The Valley derives its name from Pishin City (Map 3.8), which is the largest settlement in this area with a population of 20,479 (PCO 1998).

The Valley is connected by rail and road to Quetta (to the south) and Chaman (to the north) cities. The city of Quetta is the provincial headquarters of Balochistan, and is about 55 km away from the Valley's center (Map 3.1). Chaman is a small city, nearly 100 km away from Quetta in the north-west. It is a very busy market for smuggled goods due to its location on the Pak-Afghan border. The Pishin River flows from northeast to southwest, almost through the center of the long axis of the Valley (Fig. 3.3). The total area of the Valley is 2,416 km$^2$ (derived from the GIS outline map of the Valley) (Map 3.2).

A.S. Khattak, *Mutual Sustainability of Tubewell Farming and Aquifers: Perspectives from Balochistan, Pakistan*, Advances in Asian Human-Environmental Research, DOI 10.1007/978-3-319-02804-0_3, 

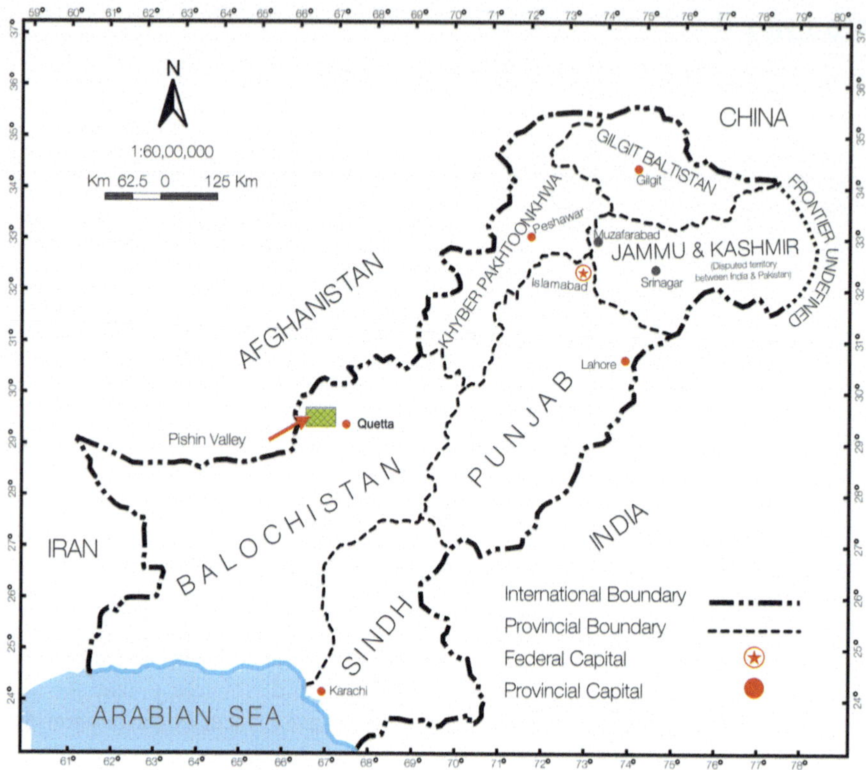


**Map 3.1** Location of Pishin Valley in the country's geographic frame

In the current administrative set-up, the Valley falls in two administrative districts of Balochistan Province, which are as follows:

1. Q. Abdullah (also spelled as Killa Abdullah) District covers 1,568 km$^2$ of the Valley's land (Fig. 3.2). Total area of the district is 3,293 km$^2$.
2. Pishin District covers 848 km$^2$ of the Valley's land. Total area of the district is 7,819 km$^2$.

Figure 3.1 reveals that the Q. Abdullah district accounts for most of the Valley's land (65 %). Figure 3.2 shows that the total area of Q. Abdullah District is divided almost 50/50 between valley and non-valley. In comparison, the Pishin District contributes only 10.9 % of its total area to the valley.

Given the above facts, we have based the Valley's general scenario only on the statistics for Q. Abdullah District in terms of certain physical, social, and economic conditions. In fact, it was more logical to assume that the statistics for Q. Abdullah District were in close correspondence to the Valley's general scenario than those for the Pishin District, as the latter shares a lesser area in the Valley (Map 3.3).

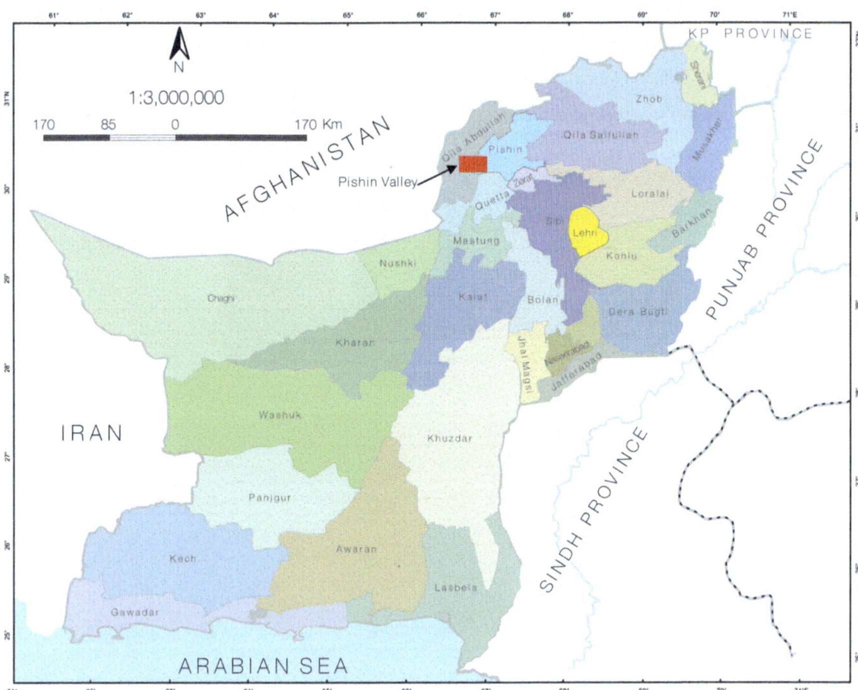


**Map 3.2** Location of Pishin Valley in the geographic frame of Balochistan province

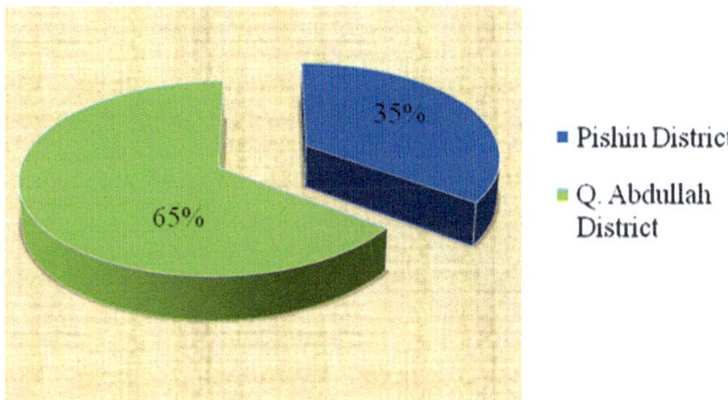


**Fig. 3.1** Area-wise distribution of Pishin Valley in Q. Abdullah and Pishin districts (Source: GIS measurement)

## 3.2 Relief

Relief influences agriculture through altitude and slope. The average altitude of Pishin Valley is 1,515 m (5,000 ft) above sea level (the elevation of Pishin City is 1,570 m (WAPDA 1993)). Geologically, the Valley is part of the Pishin sub-basin,

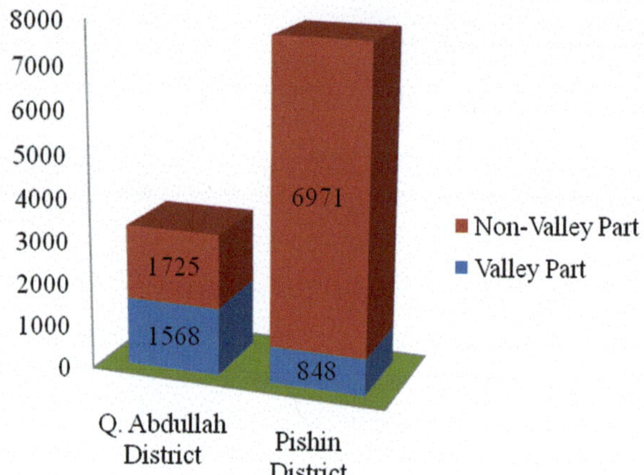


**Fig. 3.2** Area-wise distribution of Q. Abdullah and Pishin districts into their valley and non-valley parts ($km^2$) (Source: GIS measurement)

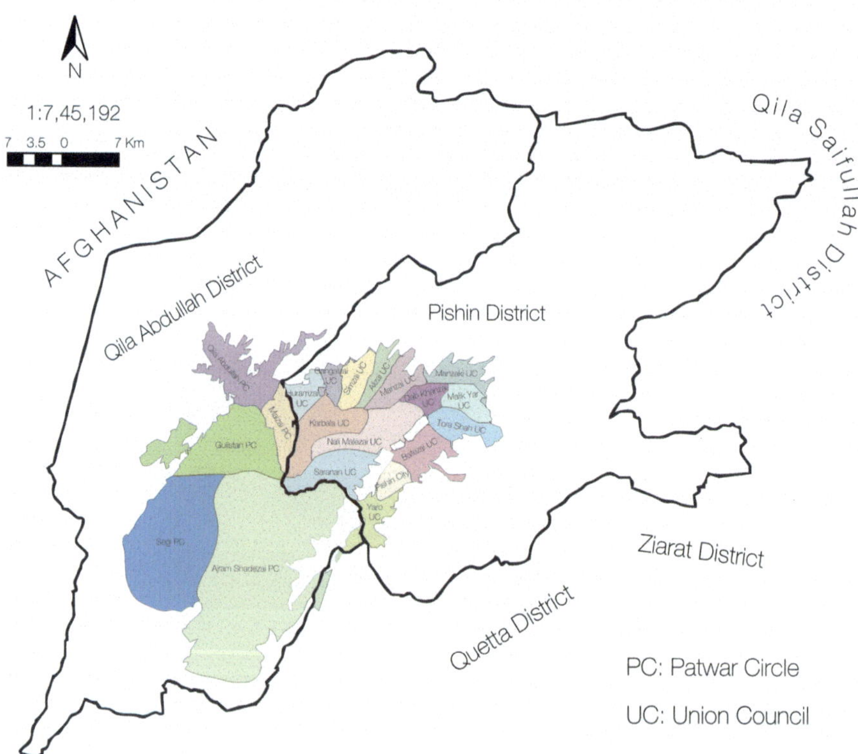


**Map 3.3** Spread of Pishin Valley in the territories of local administrative districts

the total catchment area of which is 5,193 km$^2$ (WAPDA 1993). In turn, the Pishin sub-basin is part of the grand Pishin Lora (River) Basin.

Pishin Valley can be further sub-divided into the following three geomorphological regions:

### 3.2.1 *Piedmont Plains*

These start from the toe of the surrounding mountains and are distinguished by an abrupt change in topographic gradient. The slopes are generally smooth, flat to undulating. Erosion gullies have developed in some places. The piedmonts surround the valley floor from almost all around, covering a total area of 500 km$^2$ (Map 3.4). The mountains surrounding the Valley are Sher Ghundi (northeast), Arambi (north), Khwaja Amran (northwest), Toba Kakar (west), and Ajram Ghar (south of the Valley). The piedmonts are covered with coarser material, including gravel, channel deposits, and sandy alluvium (termed as sub-recent alluvium in Map 3.4) and they are the main recharging zones.

### 3.2.2 *Valley Floor*

This is the central part of the valley characterized by low relief and graded topography. It covers an area of about 1,916 km$^2$ (calculated from Map 3.4). The floor is generally the least developed part of the valley, apparently due to its sub-standard soil, brackish groundwater, and accumulation of evaporates on the surface.

### 3.2.3 *Intra-valley Hills*

Near the southeastern and eastern frontiers of the valley lies a narrow arm of low hills, called Adam Ghar, which start near village Tirkho (30° 20′ N, 66° 47′ E) in the southeast and extend up to BKK (30° 41′ N, 67° 03′ E) in the north-east. Moreover, Map 3.4 and Fig. 3.3 also show a small hard-rock protuberance into the valley at its southwestern apex. This section of hard rock surrounds a small area of flat land connected to the main valley by narrow arms of inundation beds.

## 3.3 Lithology

The rocks exposed in the surrounding mountainous recharge zones, and inside the valley itself, are sedimentary. Their ages range from Permian to recent (WAPDA 1993). On the basis of hydro-geologic properties, they can be divided into the following two categories:

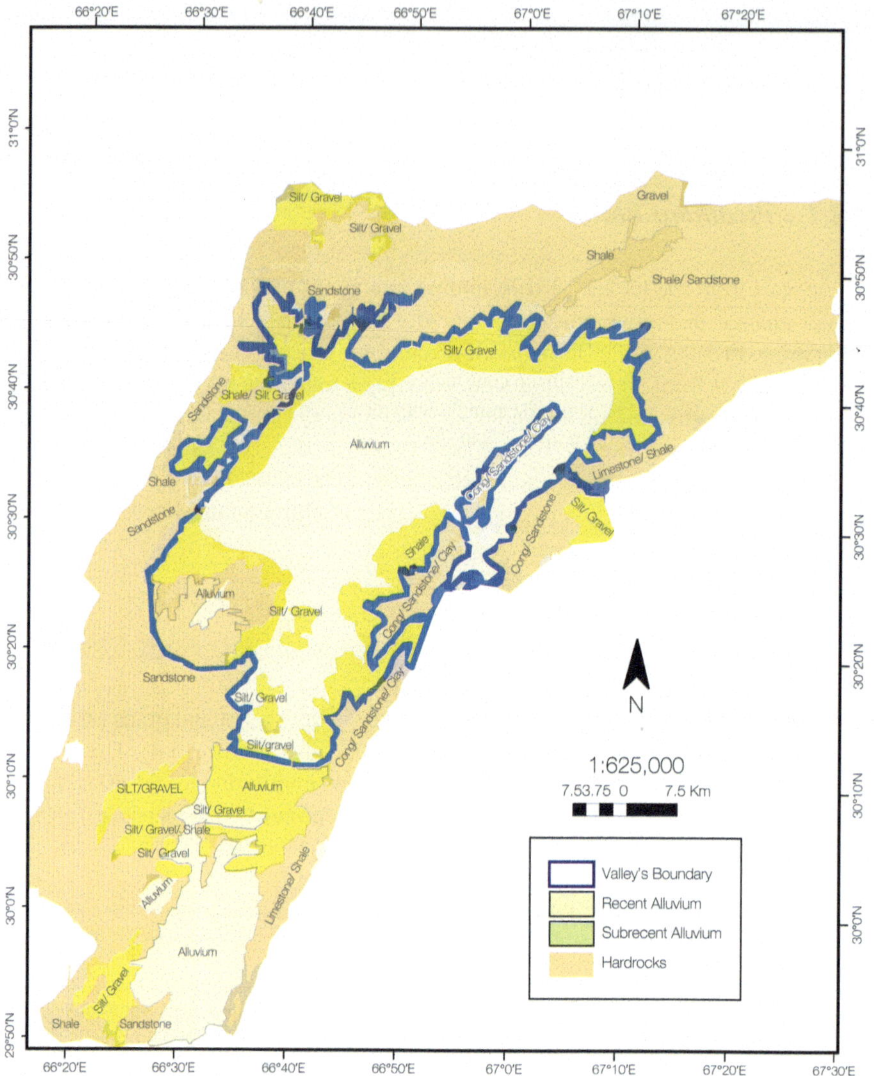


**Map 3.4** Geology of Pishin sub-basin, including Pishin Valley (Source: Geological Survey of Pakistan (GSP))

## *3.3.1 Consolidated Rocks*

The consolidated rocks of the surrounding mountains and the intra-valley hills generally behave as a barrier to groundwater movement. However, where secondary permeability has developed, they can yield enough water (WAPDA 1993). The dominant rock formations are three:

**Fig. 3.3** 3-D view of Pishin Valley and surrounding hills (Source: Landsat)

(a) Shirinab Formation is composed of thinly bedded limestone and shale. The limestone can absorb a great amount of precipitation and thus conserves it in underground storage. In contrast, the shale-bearing areas can absorb little or none at all and lose the rainwater as surface flow.
(b) Murgha Faqirzai shale and Shaigalu sandstone cover most of the mountains within the Pishin sub-basin.
(c) Bostan Formation is exposed in the southeastern watershed of the sub-basin and along Ajram Ghar, extending northeast up to BKK. The formation is composed of soft and crumbly clay and silt.

### 3.3.2 *Unconsolidated Rocks*

These are alluvial deposits constituting the piedmont plain and valley floor. They can be divided in two on the basis of their water-bearing properties:

(a) Sub-recent deposits mainly cover the foothills of the valley and are generally composed of gravel, sand, and silt. These sediments have been subjected to post-depositional changes, such as compaction and cementation. Owing to these changes, the porosity and permeability of these deposits have decreased, resulting in comparatively low water yield capacity.
(b) Recent deposits cover most of the alluvial area and constitute the main groundwater reservoir of the valley. Lithologically, these consist of silt, clay, sand, and gravel that un-conformably overlie the sub-recent deposits. Their thickness gradually increases toward the valley's axis, where it is generally over 700 ft (WAPDA 1993).

## 3.4 Soil

Depth, texture, acidity, and nutrient composition of the soil determine the types of crops that can be grown. The soils of the valley are mostly calcareous, and are derived from limestone, sandstone, and several other sedimentary rocks. These soils have a homogeneous, porous, and coarse texture; have low organic matter; and fall under over-grazed, eroded, and degraded arid lands. Mostly, they are loamy and sandy clays, as well as gravely in nature. The coarseness in texture increases toward the surrounding mountain ranges, where erosion rates are very high. The soils are distinguished by their colors, which reflect the nature of the parent material. For instance, the soils derived from limestone are brownish; those derived from shale and sandstone are greenish; and those derived from red clays, silts, sandstones, and conglomerates are brown with a reddish tinge.

Table 3.1 shows that the total area of the soils, which have very good to moderate agricultural potential (s. no. 1–3) is 109,635 acres. These three soil associations are located close to the surrounding piedmonts, which are in fact densely cultivated with fruit orchards. Some of the remaining land is rain-fed, but the majority is used as rangeland (Map 3.5 and Table 3.1).

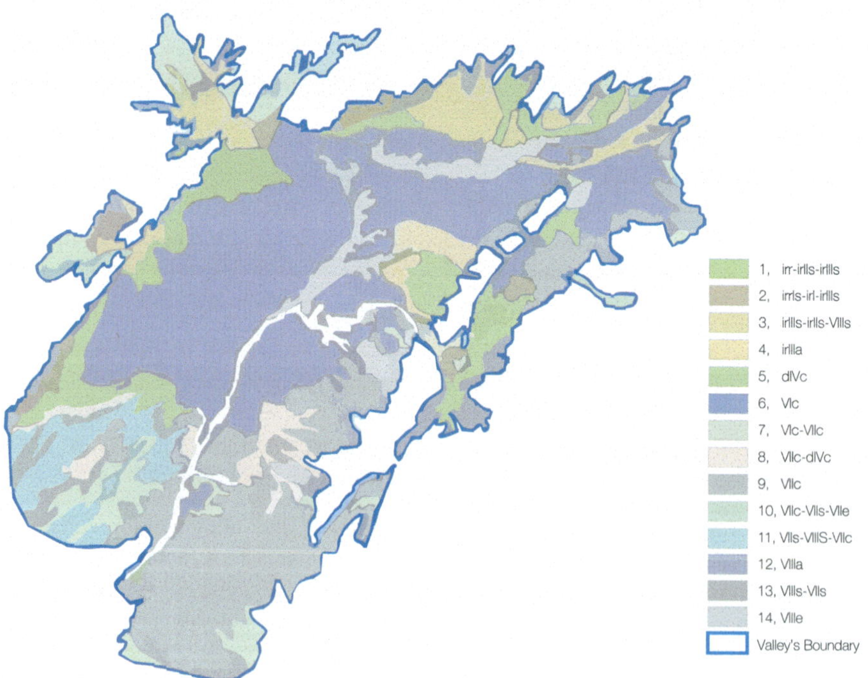


**Map 3.5** Major soil types in Pishin Valley (Source: Soil Survey of Pakistan (SSOP) 1978)

**Table 3.1** Major soil formations of Pishin Valley and their agricultural potential

| Soil formations | Area (acre) | Agricultural potential |
|---|---|---|
| irI-irIIs-irIIIs | 62,666 | Very good, with some good and moderate irrigated and irrigable land |
| irIIIs-irIIs-VIIIs | 37,211 | Moderate, with some good irrigated and irrigable land and some agriculturally unproductive land |
| irIIIa | 9,758 | Moderate irrigated and irrigable land |
| dIVc | 15,328 | Poor dry-farmed land |
| VIc-VIIc | 12,209 | Land with fair to poor grazing potentials |
| VIIc-dIVc | 14,856 | Land with poor grazing potential and some poor dry-farmed land |
| VIIc | 139,663 | Land with poor grazing potential (gravelly fans and aprons) |
| VIIc-VIIs-VIIe | 29,153 | Land with poor grazing potential (rough broken land) |
| VIIs-VIIIs-VIIc | 32,830 | Land with poor grazing potential (rocky mountains with thin soil cover) |
| VIIIa | 202,388 | Land with very poor grazing potential (saline alkali soils of the playas) |
| VIIIs-VIIs | 388 | Land with very poor grazing potential (mountains with mainly bare rocks) |
| VIIIe | 29,173 | Agriculturally unproductive land |
| Stream Bed | 11,290 | Not available for vegetative use |

Source: Statistics derived from Map 3.5

## 3.5 Climate

Climate is the main physical variable affecting agriculture. Plants and livestock require specific temperature and moisture/water conditions. Two major types of climates influence Balochistan Province as a whole:

### *3.5.1 Monsoon Climate*

High summer season (May–October) temperature and rains characterize this type of climate. During this period, moist winds rise from the Indian Ocean and Arabian Sea, which, after being elevated and diverted by the Himalayan mountains, cause usually heavy rains in most of Pakistan. The monsoonal effect diminishes from the extreme north and northeast of the country toward its south and southwest. Apart from the extreme northeastern districts of Balochistan (Zhob, Musa Khel, etc.) (Map 3.2), the rest of the province, including the study area, is rarely influenced by these winds.

### 3.5.2 Mediterranean Climate

Mild summer temperatures and winter rains characterize this type of climate. The study area bears these conditions and hence is included in the Mediterranean type of climate (Table 3.2). The climate of Pishin Valley is generally dry with average annual precipitation 6.2 in. (Map 3.7) and on the basis of average annual temperature it falls in the cool temperate category (Map 3.6). Table 3.2 gives 10 years average maximum (summer season) and minimum (winter season) temperature (40.4 °C [104 °F] and –4.8 °C [23.4 °F], respectively).

Generally, although the direct sun is troublesome due to high altitude, the shade temperature in the summer season remains averagely moderate. However, even over the highest mountains of the sub-basin, there is no perennial snow cover or glacier.

In years with normal precipitation, the annual amount ranges from 200 to 350 mm (IUCN 2006). Table 3.2 represents the acute drought spell that lasted from 1998 to 2004. It shows the average annual precipitation as 6.2 in. (165 mm). About 70 % of this amount falls in the winter season, from December to March, as both rain and snow. Winter precipitation is generally steady and widespread. Snow accumulates on soil during the winter season, which greatly contributes to the groundwater budget. A normal annual precipitation of less than 8 in. characterizes this region as a desert or semi-desert area.

Extreme variability in occurrence makes the already meager precipitation somewhat more ineffectual; and also complicates the problem of its harvesting or conservation. Due to fast wind movement and lower air humidity, potential evaporation greatly exceeds annual precipitation, resulting in the onset of desertification. In normal climatic years, the annual evaporation rates range from 2,000 to over 3,500 mm (IUCN 2006), thus necessitating irrigation for agriculture (Fig. 3.4).

The average annual evaporation is 144 in. (3,658 mm), which means that the ratio between precipitation and evaporation in the valley is 1:24. The average evapotranspiration (ETo) varies from 5.50 to 6.50 mm/day (IUCN 2006). From these figures, the value of annual ETo is derived as 2,190 mm.

## 3.6 Natural Flora

Natural flora is a good indicator of the climate of a region. The total vegetation of a region – both natural and cultivated – determines the amount of water-loss through ETo. The Pishin Valley, being arid, has sporadic natural greenery, which is also quite diverse. The natural flora mainly comprises xerophytes, which are adapted to the extreme seasonal temperature and moisture changes. Physionomically, the green cover can be classified into three growth forms:

(a) *Ephemeral Annuals*, which include dwarf grasses and short-leaved forbs, the growth activity of which is restricted to the brief moist period of 4–8 weeks around the month of March each year. The species *Bromus tectorum*, *Bromus*

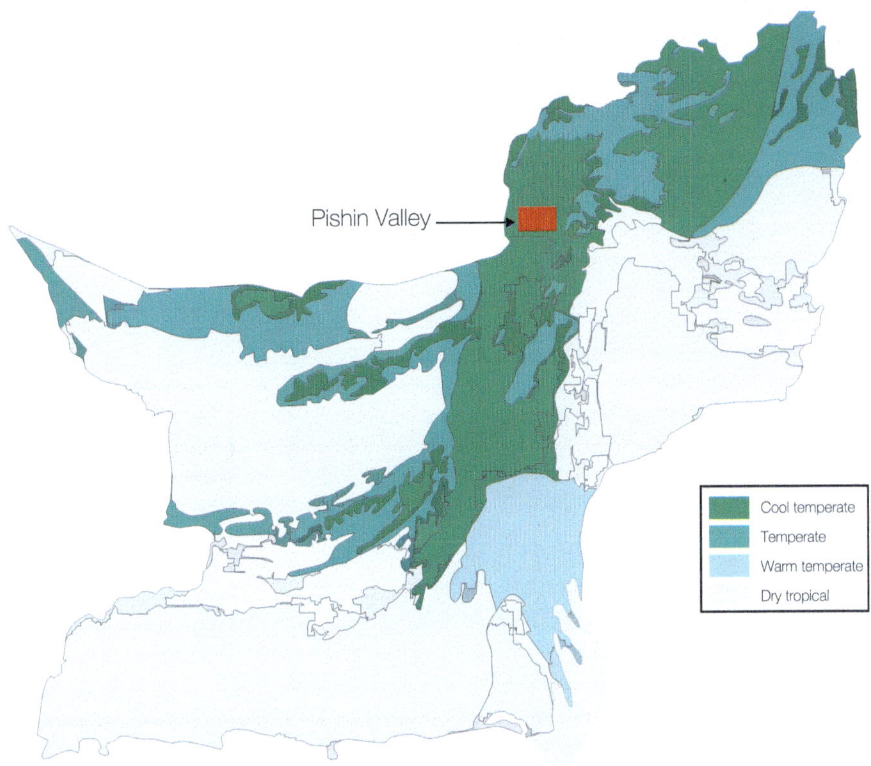


**Map 3.6** Temperature regions of Balochistan province (Source: Directorate of Water Resources Planning, Monitoring, and Development (DWRPMD), Quetta)

**Table 3.2** 10-year average of temperature, precipitation, and potential evaporation at Q. Abdullah met-station

| Years | 1996 | 1997 | 1998 | 1999 | 2000 | 2001 | 2002 | 2003 | 2004 | 2005 | G. Avg. |
|---|---|---|---|---|---|---|---|---|---|---|---|
| Temp. max. °C | 37 | 39 | 42 | 40 | 41 | 42 | 42 | 40 | 39 | 42 | 40.4 °C<br>104 °F |
| Temp. min. °C | −6 | −4 | 0 | −5 | −5 | −7 | −6 | −6 | −3 | −6 | −4.8 °C<br>23.4 °F |
| ppt. (in.) | 4.72 | 11.01 | 7.50 | 7.06 | 3.38 | 2.74 | 4.99 | 4.15 | 5.02 | 11.54 | 6.2 in.<br>165 mm |
| P. evap (in.) | 143.7 | 131.3 | 139.1 | 150.5 | 151.2 | 136.1 | 142.2 | 149.1 | 159.7 | 136.5 | 144 in.<br>3,658 mm |

Source: DWRPMD, Quetta
Note: Q. Abdullah Town and its weather station are located within Pishin Valley

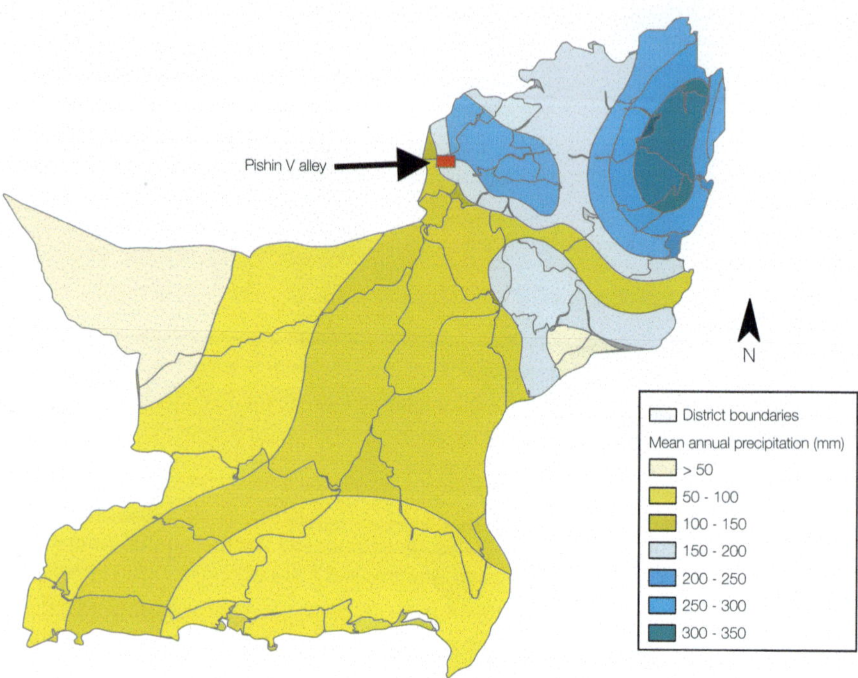


**Map 3.7** Precipitation regions of Balochistan province (Source: Directorate of Water Resources Planning, Monitoring, and Development (DWRPMD), Quetta)

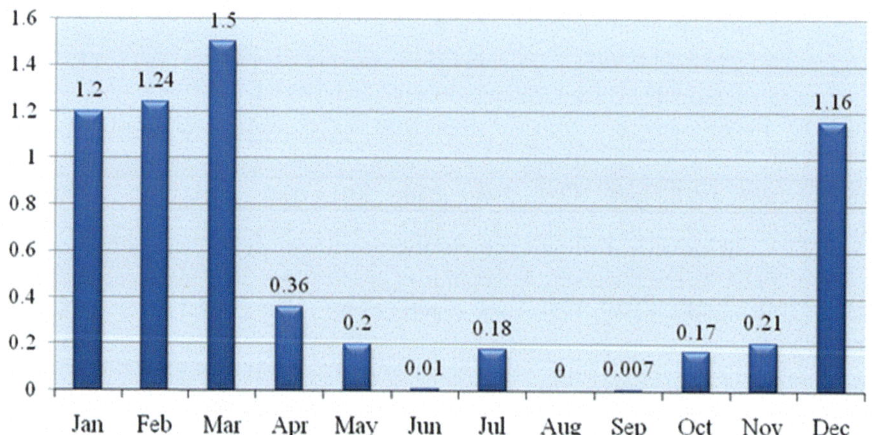


**Fig. 3.4** Monthly precipitation averages (inches) for Q. Abdullah station, 1995–2006 (Source: Directorate of Water Resources Planning, Monitoring, and Development (DWRPMD), Quetta)

*Japonicus*, *Onobrychis*, *Zizyphora*, *Cerastium*, etc. are among the important annuals found in this region.

(b) *Succulent Perennials*, which are confined to the saline piedmont slopes. They, by their structural and physiological characters, store water during the rainy season for consumption in the dry period. *Sasola* spp. and *Haloxylon recurvum* are the two commonly found succulent perennials in this area.

(c) *Non-Succulent Perennials*, which include shrubs and grasses. The hardy shrubs dominate the landscape in the rangelands. Their survival is due to their wide and deep rooting system for sucking the little soil moisture, and a short offshoot system to minimize transpiration rates. The important shrubs found here are *Artimisia maritime*, *Haloxylon griffithii*, *Caragana ambigua*, etc. Among the perennial grasses, *Chrysopogon*, *Cymbopogon*, and *Pennisetum* are the hardiest species, which can resist against complete grazing.

## 3.7 Livestock

Livestock use to have a stake in the direct and indirect use of water resources. They can modify the local ecology by grazing natural flora, and a large amount of farmland is given to forage cultivation for their feed. It is an important income generator, as well as a source of nutrition for the people of the area. Low human population and the presence of extensive rangelands are highly conducive for livestock breeding in the valley. Apart from occupational herding, small ruminants and poultry birds are reared by almost every household in every village. The mountain slopes and the gravelly piedmont fans and aprons are predominantly used as grazing lands.

In Pakistan's official record, livestock data are registered on the basis of administrative districts. From the total livestock data for Q. Abdullah and Pishin Districts, the statistics for Pishin Valley are derived per its territorial ratio in the two constituting districts (Table 3.3). Aside from poultry, sheep, goats, and cattle are very common in the valley. Since the area is deficient in modern means of transportation, some freight animals, like camel, horse, ass, etc. are also reared. In terms of density per sq. kilometer, sheep are dominant at 61 heads, followed by poultry (60 heads), goat (38 heads), and cattle (11 heads). Such a high livestock density has a conspicuous impact on the arid ecology of the valley.

Grazing is generally of nomadic type. Traditionally, every year a large number of herding nomads cross the Afghanistan border into the Pishin Valley whenever the season becomes harsh there (transhumance still continues between the two countries). Their herds mostly comprise goats, sheep, and camels. They particularly keep freight animals, as well as dogs for guarding. The grazing is uncontrolled and there is no range management such as planned rotational grazing to allow re-growth of the palatable vegetation.

**Table 3.3** Livestock population

| Area | Livestock | | | | | | | | | |
|---|---|---|---|---|---|---|---|---|---|---|
| | Cattle | Buffalo | Sheep | Goat | Camel | Horse | Mule | Ass | Poultry | Total |
| Q. Abdullah total | 41,966 | 192 | 186,504 | 99,015 | 375 | 393 | 96 | 3,219 | 224,983 | 556,743 |
| Pishin Valley share (@ 47.6 %)[a] | 19,976 | 91 | 88,776 | 47,131 | 178 | 187 | 46 | 1,532 | 107,092 | 265,009 |
| Pishin Total | 69,991 | 1,928 | 557,089 | 434,507 | 2,400 | 3,637 | 557 | 15,740 | 363,869 | 1,449,668 |
| Pishin Valley share (@ 10.5 %)[a] | 7,349 | 202 | 58,494 | 45,623 | 252 | 382 | 58 | 1,653 | 38,206 | 152,219 |
| Pishin Valley Total[a] | 27,335 | 293 | 147,270 | 92,754 | 430 | 569 | 104 | 3,185 | 145,298 | 417,228 |
| Density/km$^{2}$[a] Pishin Valley | 11 | 0.12 | 61 | 38 | 0.18 | 0.24 | 0.04 | 1.3 | 60 | 173 |

Source: Livestock and Dairy Development Department, GoB

[a]Derived statistics

## 3.8 Demography

### *3.8.1 Population Size and Distribution*

Since the Pishin Valley is a physical region, rather than an administrative unit, we were unable to find summarized population statistics for its specific territorial limits during our survey with any of the concerned public sector departments. However, since the valley constitutes about the half of Q. Abdullah District (Fig. 3.2), its population structure is represented by this district, which registered a total of 370,269 people in the 1998 census. The district's population was 176,341 in the census of 1981. Thus, it recorded an annual increase of 4.46 % over the 17-year intercensal period. Male to female ratio in the district is 121.4 (i.e. for every 100 females, the males are 121.4) rural proportion is 84.7 %, and the average household size is eight individuals (PCO 1998). Population densities in Q. Abdullah and Pishin Districts are 112.4 and 47 individuals per sq. km, respectively (PCO 1998). Pishin Valley, being a plains area, obviously has the relatively more dense rural units of these two districts.

Map 3.8 shows that the valley is uniformly dotted with settlements. However, the larger population units are located at/or near the piedmont zones because these areas have better quality soil and sufficient water resources. Table 5.1 provides primary data on some of the population characteristics. It reveals that a joint-family system prevails among the farming communities of the valley, with an average household size of 31.7 individuals.

### *3.8.2 Literacy and Education*

The 1998 population census of Pakistan defines a literate person as one who can read a newspaper and write a simple letter in any language. Per this definition, the literacy ratio in the rural Q. Abdullah District is only 12 % among the population aged 10 years and above. The same census defines an educated person as one who attains at least a primary level of formal education. Of the total educated individuals (urban and rural) in the district, 35.8 % have a primary level of education (5 years of education), 22.5 % middle (8 years), 12.8 % matriculation, 2.3 % intermediate, 1 % graduation (14 years), and 0.53 % post-graduate education. Overall, male and female literacy ratio is 6:1. Female education at higher levels is very rare. Apart from formal education, a great number of students also seek knowledge from religious schools (madrasas), which operate on a welfare basis.

Unfortunately, 33.3 % of the valley's farmers are completely illiterate, while about 60 % of the remaining are educated up to primary and high school levels only (Table 5.1). This is one of the major reasons that it may be difficult to achieve modernization of the agricultural sector in the near future.

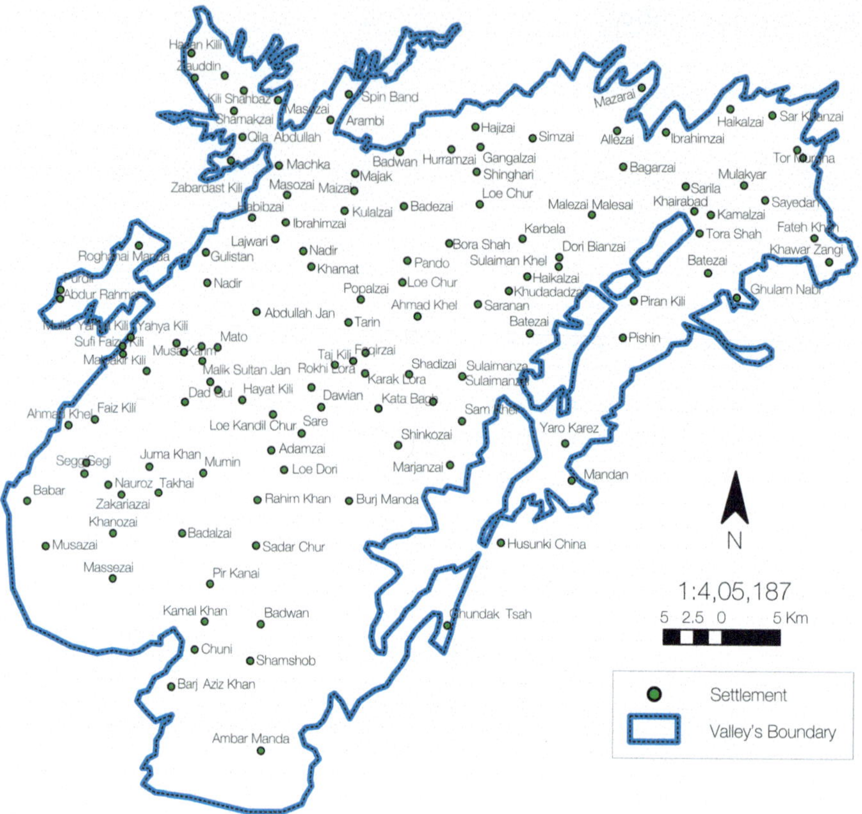


**Map 3.8** Distribution of settlements in Pishin Valley (Source: Topographical sheets of the Survey of Pakistan (SOP))

### *3.8.3 Occupations*

Cultivation farming and livestock herding are the major occupations. An average household spares 3.6 of its adult males for farming activities (Table 5.1). The women help their men-folk in crop harvesting and threshing. Children normally rear herds in the nearby grazing grounds, called rangelands. Other sources of income include government jobs, trade, and commerce. The business activities of the valley's inhabitants are widespread in the country and abroad. In fact, the prosperity of Quetta City in trade, commerce, and transport is mainly owing to the investments made by the locals of Pishin Valley. The area interlinks trade flow between Pakistan and Afghanistan through the Quetta-Chaman road and railways.

As secondary data on occupational structure in the study area were unavailable, our survey included questions relating to this (Table 3.3). Only 31.5 % of households can make their whole living from their farms; the remaining farmers must also complement their farm income with some non-farm businesses in order to live respectably despite the human–land ratio in the valley is generally high.

### 3.8.4 Socio-political Organization

The valley is mostly inhabited by Pashtoons (Also called 'Pathans' in Urdu language). Although their sub-tribes are many, there is no noticeable cultural variation among them. Decision-making is usually centralized. A chieftain, who usually also becomes an elected representative in the state political bodies, represents each sub-tribe. Normally, tribes comply with the advice and directives of their chieftains or local elders. These notables are approached for resolution of social conflicts of every kind, and the decisions so taken are honored strictly. In particular, the conflicts concerning irrigation water are resolved through community-based reconciliatory bodies instead of court proceedings. Due to harmonious cultural traits, customs of mutual cooperation prevail among the society. In times of need, the labor force is pooled (called 'ashar') and utilized reciprocally for one another. This is often seen during sowing and harvesting periods. The Government has thus far been unable to withdraw the 'flat rate policy' in electricity tariff for agricultural tubewells, despite it being thought of as detrimental to the meager groundwater resources. This is basically because of the overwhelming political force of the landlords, who are in favor of this policy. In fact, in the past, several political regimes have announced withdrawal of the flat rate policy after recommendation by expert bureaucracy; however, the regimes have soon surrendered to the resentment of the so-called Farmers' Action Committees.

## 3.9 Irrigation System

Due to scanty precipitation and high rates of ETo, a paying agriculture in Pishin Valley is only possible through irrigation. The surface water resources are meager and unreliable; hence, groundwater is almost the exclusive source of irrigation. Presently, there are three main systems of irrigation in practice: canals, karezes, and tubewells.

### 3.9.1 Canal System

Since the British Rule over the area, a few public irrigation projects have been built and operated by public sector departments. Principal among them are the following:

### BKK

BKK is a historical source of surface water irrigation. It was built in 1885 to create storage for the floodwaters of Tor Murgha Nala and Barshor River, together with limited high-water diversion from Surkhab River. The structure consists of a 'bund' (earth dam or dike), which is nearly a half mile long and 50 ft high. The original capacity of the reservoir was about 40,000 acre Ft., which has now been reduced by half by sedimentation and illegal encroachments.

### Malezai Pumping Plant

The Malezai pumping plant is built on the Pishin Lora stream, which is formed by the confluence of the Barshore and Tor Murgha streams, about 5 miles downstream from the Tor Murgha Headworks. At that point, a perennial flow of 8–10 cfs is provided to delivery canals that serve the lands on the southern flank of the stream as a supplement to the service from BKK (Map 3.9).

### Shebo Headworks

The Shebo headworks is a diversion canal from Kakar Lora to the same area served by the aforementioned two installations.

### Pishin Canal

When flow is available, the Pishin Canal diverts the water of the Surkhab Stream to the area surrounding the Pishin town (Map 3.10).

## 3.9.2 *Karez System*

A karez (known as 'Qanat' in Iran) is a long (usually 1–2 km) system of underground tunnels provided with vertical shafts at regular distances for ventilation. The tunnels are laid on a natural gradient to convey water from a mother well located at the foothills to a daylight point from where the water is finally distributed in open channels for multiple uses, including domestic, livestock, and agriculture. The karezes are community-based systems, and a flowing karez has to be cleaned almost every year to keep them functional. Cleaning a karez requires much skill, courage, and labor; thus, the users bear enormous costs on a recurring basis almost every year.

Due to the inherent meagerness and unreliability of surface water resources, karez irrigation has been a tradition in the study area since the start of the

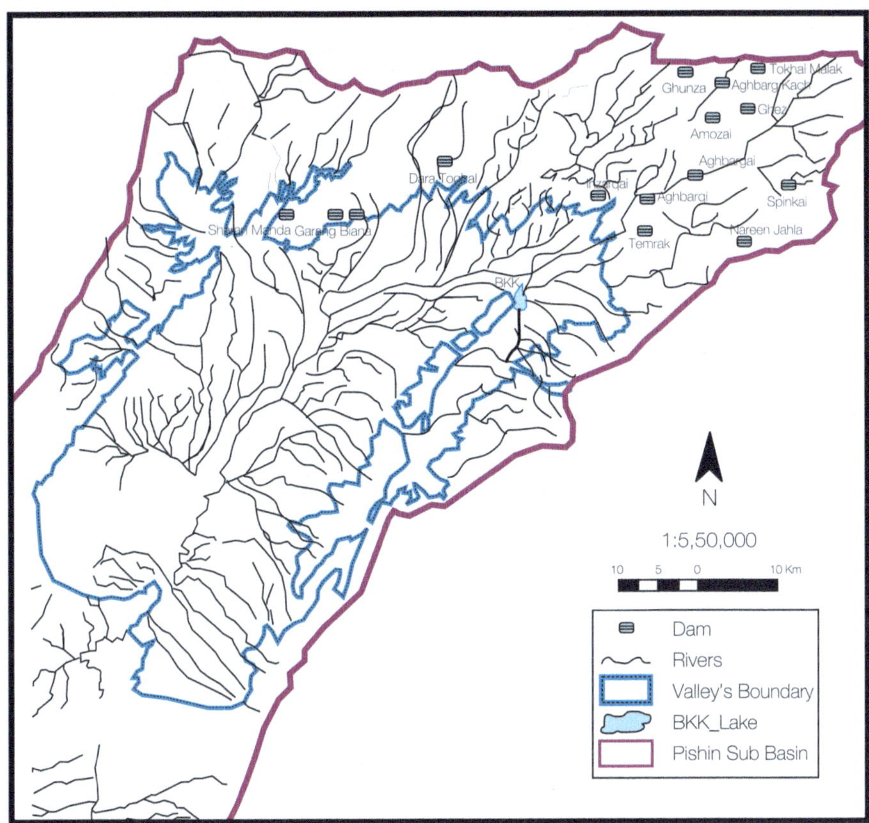


**Map 3.9** Distribution of DADs in Pishin sub-basin (Source: Irrigation Department, Government of Balochistan (GoB))

nineteenth century. However, with the electrification of the area, this system gradually disrupted, giving way to individual-based tubewell systems. Other reasons for its failure include the enormous fall of the watertable, common source tragedies, and electricity tariff subsidies for irrigation tubewells, etc. Hundreds of karezes were once functional in the area; however, very few of them are currently giving service.

### *3.9.3 Tubewells*

Subsoil water was traditionally tapped from natural springs or through karezes. However, in recent years, tubewells have become more common and are mostly used for growing fruits and vegetables, and to supplement other sources of irrigation. The area under fruit orchards, especially those of deciduous trees, has increased

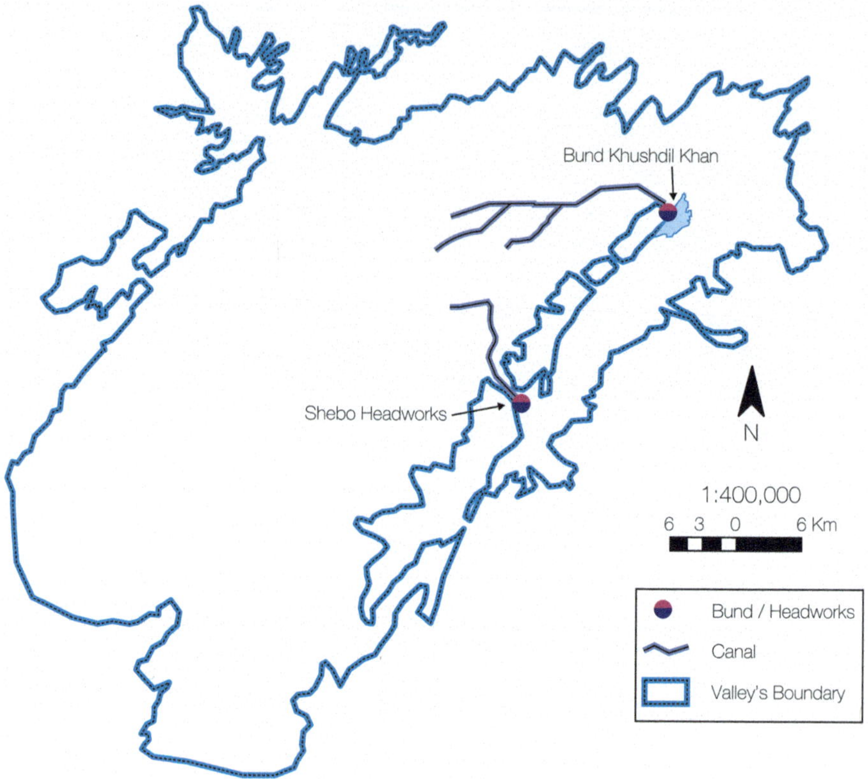


**Map 3.10** Surface water irrigation canals in Pishin Valley (Source: Irrigation Department, Government of Balochistan (GoB))

tremendously in the recent past, only through use of tubewells, and this has brought positive change in living standards (Map 3.11).

## 3.10 Farm Size and Tenure System

In Q. Abdullah District, which constitutes the bulk of Pishin Valley, the average farm size is 14.7 acres (Agriculture Census 2000). The size of the majority of farms (24 %) is in the range of 7.5 to approximately 12.5 acres. In terms of area, the class that ranges between 12.5 and 25 acres has the lead, with 23 % in the total farm area of the district (Table 3.4). There is no Government-owned farm in the district.

While pasturelands (called 'shamilat', which means common land) belong to some local tribes as a whole, the agricultural area is under personal ownership. Normally, the landowners themselves manage cultivation activities (Table 3.5).

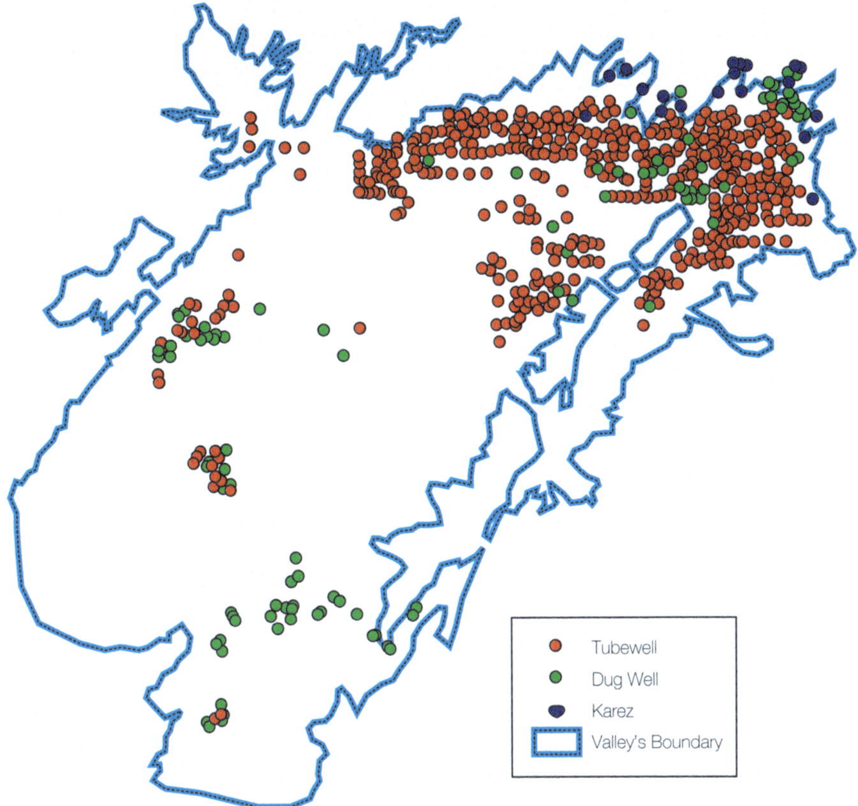


**Map 3.11** Distribution of water-yielding points (2006) (Source: Directorate of Water Resources Planning, Monitoring, and Development (DWRPMD), Quetta)

The workforce consists of family members as full- or part-time workers, permanently employed laborers, or casual daily wage earners. Where a farm is not managed by the owner himself, three types of tenancy contracts exist in the area:

1. *Tenant-Share Tenancy* – a system of tenancy where the landowner bears the full cost of inputs, including irrigation, and the produce is divided according to a set formula between the owner and a tenant (normally one-fourth for the tenant).
2. *Owner-Share Tenancy* – a system of tenancy where the tenant bears the full cost of inputs, including irrigation, and the produce is divided according to a set formula between the owner and the tenant (normally one-fifth for the owner).
3. *Cash Tenancy* – a system of tenancy where the tenant leases the land (usually a fruit orchard) for an agricultural year on a lump-sum payment basis.

The type of tenancy relates to irrigation frequency and methods. It is logical to premise that a non-owner user of a tubewell makes maximum use of the resource

**Table 3.4** Number and area of farms by size of farms, Q. Abdullah district

| | | | Farm area | |
|---|---|---|---|---|
| Farm size (acre) | Total farms | % of total farms | Total area | % of total area |
| All farms | 15,514 | – | 227,795 | – |
| Govt. farms | – | – | – | – |
| Private farms | 15,514 | 100 | 227,795 | 100 |
| <1 acre | – | – | – | – |
| 1–2.5 acres | 2,233 | 14 | 3,793 | 2 |
| 2.6–5 acres | 2,220 | 14 | 7,373 | 3 |
| 5.1–7.5 acres | 2,407 | 16 | 14,138 | 6 |
| 7.6–12.5 acres | 3,691 | 24 | 36,387 | 16 |
| 12.6–25 acres | 3,128 | 20 | 52,951 | 23 |
| 25.1–50 acres | 1,003 | 6 | 39,684 | 17 |
| 50.1–100 acres | 601 | 4 | 45,497 | 20 |
| 100.1–150 acres | 227 | 1 | 27,022 | 12 |

Source: Agriculture Census Report, 2000, Balochistan Province

**Table 3.5** Tenure classification of farms, Q. Abdullah district

| | Tenancy status (% of total farms) | | |
|---|---|---|---|
| Size of farms (acres) | Owner | Owner-cum-tenant | Tenant |
| All farms | 98 | 0.2 | 1.8 |
| <1 acre | – | – | – |
| 1–2.5 acres | 99 | – | 1 |
| 2.6–5 acres | 100 | – | – |
| 5.1–7.5 acres | 91 | – | 9 |
| 7.6–12.5 acres | 99 | – | 1 |
| 12.6–25 acres | 100 | – | – |
| 25.1–50 acres | 98 | 2 | – |
| 50.1–100 acres | 100 | – | – |
| 100.1–150 acres | 100 | 0 | – |

Source: Agriculture Census Report, 2000, Balochistan Province

without any care for its sustainability. Similarly, an ad hoc tenant would prefer the cheaper flood irrigation method to the HEISs, because the latter requires higher initial investment.

## 3.11 Cropping Pattern

Cropping pattern may be defined as 'allocation of the area of a farm to the various crops grown in a specific year in an agro-ecological zone'. The factors affecting cropping patterns include climate, soil, availability of irrigation water, marketing conditions, and social factors like technical skill of labor, population density, size of holdings, etc. Balochistan province is divided into four major agro-ecological zones.

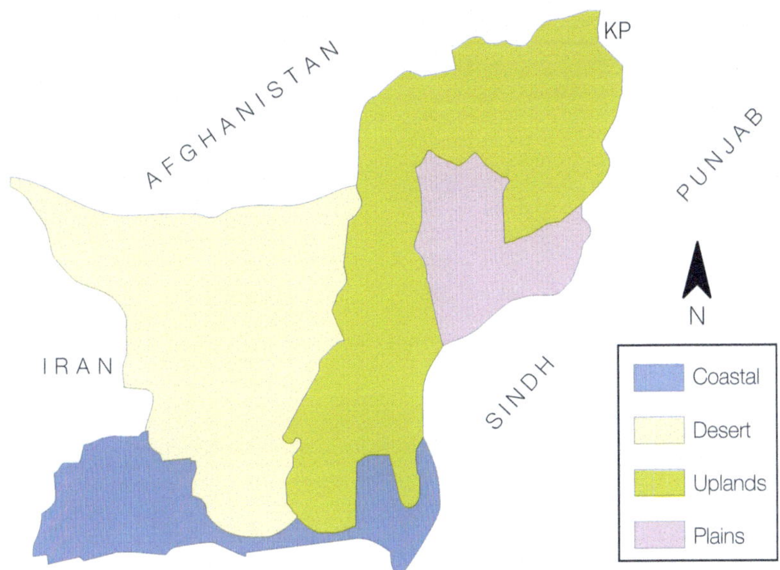


**Map 3.12** Agro-ecological zones of Balochistan province (Source: Directorate of Water Resources Planning, Monitoring, and Development (DWRPMD), Quetta)

The Pishin Valley falls in the 'uplands zone', which is characterized by cold winters (with snowfalls and frost) and mild summers (Map 3.12).

The dry temperate climate is ideal for deciduous fruits. There are two cropping seasons in the Valley:

### *3.11.1 Kharif Season*

This is the season of summer crops, which grow between mid-April and mid-October. This season is the main contributor in agricultural production, provided irrigation water is available. The crops grown during the kharif season include apples, plums, apricots, pomegranates, grapes, melons, vegetables, tobacco, potatoes, forage, and onions, etc.

### *3.11.2 Rabi Season*

This is the season of winter crops, which grow between mid-October and mid-April. This season traditionally observes a large amount of farm fallowing, because the low temperatures, frost, and snowfall restrict vegetative growth in most crops.

**Table 3.6** Area, production, and growing seasons of all crops in Q. Abdullah district

| | Area in hectares | | | Production in tons | | | | |
|---|---|---|---|---|---|---|---|---|
| Crops | Irrigated | Un-irrigated | Total | Irrigated | Un-irrigated | Total | Sowing time | Harvesting time |
| Rabi crops | | | | | | | | |
| Wheat | 1,900 | 400 | 2,300 | 3,938 | 472 | 4,410 | 15 Nov–15 Dec | 15 Jun–15 Jul |
| Barley | 135 | – | 135 | 202 | – | 202 | Dec–Jan | Jun |
| Cumin | 25 | 18 | 43 | 20 | 7 | 27 | Mar–Apr | Jun |
| Vegetables | 50 | – | 50 | 798 | – | 798 | Sep–Oct | Dec– Onward |
| Forage | 228 | – | 228 | 3,878 | – | 3,878 | Mar | May |
| Kharif crops | | | | | | | | |
| Fruits | 6,041 | – | 6,041 | 31,866 | – | 31,866 | 15 Feb–10 Mar | Jul–Nov |
| Onion | 80 | – | 80 | 1,456 | – | 1,456 | 15 Mar—Apr | Sep–Oct |
| Potato | 150 | – | 150 | 2,865 | – | 2,865 | Apr–May | Oct–Nov |
| Vegetables | 238 | – | 238 | 1,831 | – | 1,831 | Apr | Sep–Oct |
| Melons | 96 | 35 | 131 | 1,151 | 302 | 1,453 | Apr | Sep–Oct |
| Forage | 200 | – | 200 | 2,901 | – | 2,901 | Apr | Jun |
| Tobacco | 520 | – | 520 | 24,109 | – | 24,109 | May–Jun | Sep–Oct |

Source: Agriculture Statistics of Balochistan, 2004–05

However, this farm-fallowing feature is now being greatly equalized by the kharif season as well due to the water scarcity then. Important Rabi crops are wheat, barley, cumin, vegetables, and forage.

The following three cropping systems prevail in the Valley:

### Fruit-Based System

The fruit-based system refers to fruit tree cropping. The major fruits in this list are apples, grapes, and apricots. Where water is too scarce, grapes are preferred because of their low-delta property. However, the most dominant fruits are apples, followed by apricots and grapes. Other fruits are almonds, plums, pomegranates, peaches, pears, and pistachios; however, they are not grown on a commercial scale.

### Crops-Based System

The major crops grown in the crop-based system are wheat, barley, cumin, melons, vegetables, onion, chilies, potatoes, carrots, tobacco, and forage. The former four crops are also grown through dry farming. Tobacco is a recent addition due to its high market value and the ability to flourish on a saline soil (Table 3.6).

### Mixed System

As its name implies, the 'mixed system' mixes both fruit trees and field crops. It prevails in areas very close to local population centers, like Pishin Town itself, or in areas on the major roads. Horticulture is this region's traditional and

**Table 3.7** Area, production, and yield of fruits, Q. Abdullah district

| Fruits | Area in hectare | Production in tons | Yield in kg/ha |
|---|---|---|---|
| Apple | 4,529 | 23,406 | 6,830 |
| Apricot | 949 | 6,313 | 8,351 |
| Grapes | 282 | 979 | 4,618 |
| Peach | 33 | 82 | 5,857 |
| Plum | 40 | 36 | 4,000 |
| Pomegranate | 120 | 971 | 11,161 |
| Others | 25 | 79 | 3,591 |
| Total | 6,041 | 31,866 | – |

Source: Agriculture Statistics of Balochistan, 2004–05

staple cultivation (Table 3.7), and vegetables, tobacco, and other high-value/low-delta crops are mixed with it.

## 3.12 Summary

This chapter described several physical and human characteristics of the study area, which can have a stake in the area's agriculture and water resources. It revealed that, since Pishin Valley is a physical region, not an administrative unit, secondary summarized statistics for its entire territory are not available for any parameter. The valley is located close to the Pak-Afghan border and is divided between two administrative districts of Balochistan Province. It has a Mediterranean-type climate, which is semi-arid and cool temperate. Natural vegetation is sporadic and the livestock consists of herds and small ruminants. The valley is moderately populated and the rural literacy ratio is 12 % among the population aged 10 years and above. Further, the chapter revealed certain traditional characteristics, like irrigation system, land tenure system, cropping pattern, etc., so that the prevailing practices in resource utilization and management can be understood.

## References

Agriculture Census Report (2000) Statistics Division, Government of Balochistan, Quetta

IUCN (2006) Water requirements of major crops for different agro-climatic zones of Balochistan. The World Conservation Union (IUCN) Pakistan, Water Programme. Available at: http://cms-data.iucn.org/downloads/pk_water_req.pdf, 20 July 2013

Population Census Organisation (1998) Population Census Report, Pakistan

WAPDA (1993) Groundwater levels during 1976–93 in Pishin Sub-Basin, Basic Data release no. 4, Hydrogeology project. WAPDA, Balochistan (unpublished)

# Chapter 4
# Material and Methods

**Abstract** This study analyses the interdependent changes in tubewell irrigated agriculture and aquifer potential within a time frame of 28 years, starting from 1981. The reason for tracking this dual issue from 1981 is rooted in the environment created by the military occupation of Afghanistan by the former USSR in 1979. Apparently, to counter and check the socialist philosophy in the region, the USA made Pakistan its frontline ally and provided financial aid/loans to her over the next couple of years. A US-sponsored NGO, USAID, was very active in this regard. Besides her, some other wealthy countries of the US camp also assisted Pakistan in various development projects in the early 1980s. Thus, that was the peak time for the study area to become widely electrified. This brought a boom in tubewells; and thus the enormous exploitation of the aquifer set in. This chapter discusses the material used and the methodology applied in the study at hand. The research is based on questionnaire surveys and satellite images. Besides time-series data of agricultural land use, watertable, water quality, and farm income, the point data of certain variables, like cropping pattern, institutional role, etc. has also been used. This chapter entails the types and sources of data, the history and strengths of Landsat data, sampling procedure, survey tools, etc.; and the software and techniques used for data processing and analysis.

**Keywords** Methodology • Farmers • Survey • Satellite images • Analysis

## 4.1 Framework of Methodology

Figure 4.1 summarizes the methodology applied in this research. The flow of activities begins with taking special learning courses over a period of 1 year (24 credit hours) at the host university – the Punjab University, Lahore. After that, synopsis of the research was developed through review of the related literature, site selection, and reconnaissance. The archival sources of information have helped in comprehending

A.S. Khattak, *Mutual Sustainability of Tubewell Farming and Aquifers: Perspectives from Balochistan, Pakistan*, Advances in Asian Human-Environmental Research, DOI 10.1007/978-3-319-02804-0_4, 

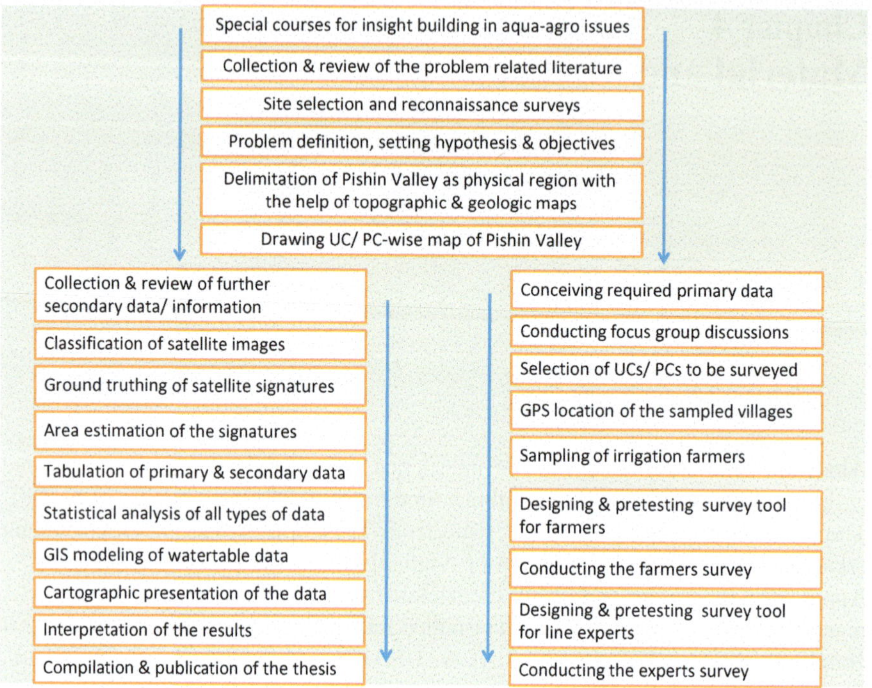


**Fig. 4.1** Flow chart of the applied methodology (Source: Saeed 2012)

the problem to be studied, delimitation of the study area, and in selection and design of the data collection tools.

## 4.2 Selection of Study Area

The major criteria that justified selection of the Pishin Valley for this nature of study are explained hereunder:

- The area is a plain landscape of rural character that possesses vast pockets of historically intensive irrigated agriculture. The area's groundwater economy was reported, as well as directly observed by the author himself, to have come under stress, apparently due to social and institutional deficiencies.
- The area is entirely within the Pishin sub-basin of the Pishin River. Therefore, analysis of its ecological resources was in concordance with the 'integrated basinal approach' that has been strongly recommended by most of the related literature.
- The entire area is conveniently accessible due to a good internal communication infrastructure, as well as its proximity to Quetta City, which is the provincial headquarters and, thus, the home of almost all concerned departments.

- People of the area are generally well educated and better civilized than the rest of the people of Balochistan. They were expected to willingly and sensibly participate in the interviews.
- The local community bears the same ethnic and linguistic features as the author himself. Therefore, a good rapport and mutual understanding with them was likely.
- The author, being a university teacher, had a number of students who were native inhabitants of the study area; hence, their expected assistance in the field survey was a considerable bonus.

## 4.3 Data Types and Sources

This study used both secondary and primary data. Details are given hereunder:

### *4.3.1 Secondary Data*

The main types of secondary data include statistical tables, maps, and satellite images (Fig. 4.2); detail is given hereunder:

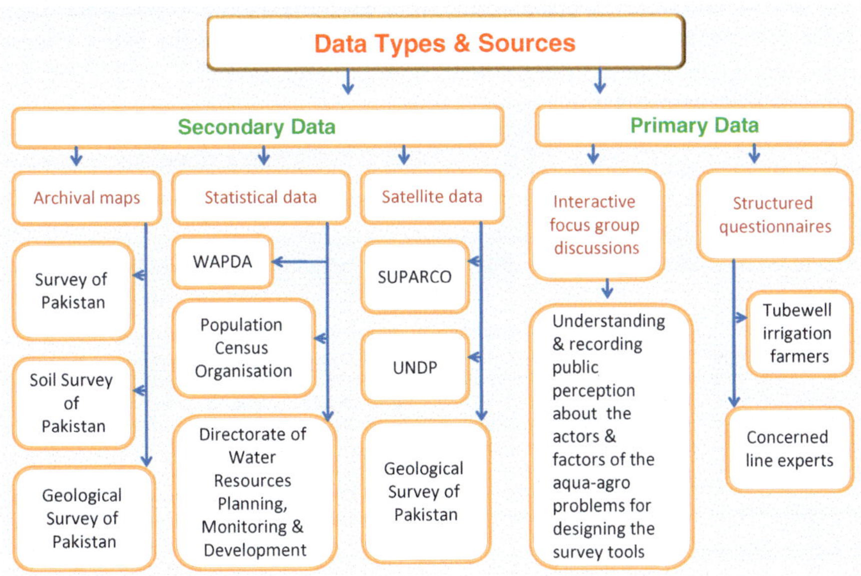


**Fig. 4.2** Scheme chart of the type of data used and its sources (Source: Saeed 2012)

### Statistical Data

Archival climatic data, recorded at the Q. Abdullah town weather station (located 1,589 m above MSL), were collected from the DWRPMD, Balochistan. Moreover, the work of WAPDA, DWRPMD, the Irrigation & Power Department, and some private organizations such as Halcrow Rural Management Consultants, Cameos, etc., in the fields of assessment of aquifer potential, discharge rate of local streams, development of check dams and the minor canal irrigation schemes (e.g. BKK and Shebo Headworks), growth of water-yielding points (dugwells/tubewells/karezes), adoption of micro-irrigation technologies, etc. were used. Statistics regarding farm size, crops, and livestock were extracted from the Agricultural Census Reports for Balochistan, and the provincial Livestock & Dairy Development Department. Population and literacy data were sourced from the PCO of Pakistan.

### Locational Data

As the year 1981 was the reference point for tracking aqua-agro changes, the agricultural land use scenario of that time was delineated from a survey of Pakistan's topographical sheets. Although the majority of the toposheets that we were able to collect (possession of toposheets is legally restricted in Pakistan) were published during the 1960s; however, going by the logic that land use changes prior to 1980 were slow paced, we projected those data for the 1981 scenario. The total 2,416 sq. km area of the valley accommodates 11 toposheets of 1: 50,000 scale each (Table 4.1). Several of the sheets cover a little of the area in the Valley fringes (Appendix A), but four of them (S. no. 4, 5, 8, 11 in the Table 4.1) cover the major and agriculturally populous part of the valley.

The utilized archival thematic maps include geologic maps of the entire Pishin sub-basin prepared by GSP, soil maps of the target area prepared by the Soil SOP, maps of climatic regions and ecological divisions prepared by the DWRPMD, Balochistan, etc. Some of the maps enlisted above were redrawn for direct simple presentation, while some were digitized for analysis in the Arcview GIS software.

### Satellite Data

Satellite data are the modern and most credible source of information for understanding and monitoring the behavior and distribution of surface resources and surface phenomena. Satellite images can provide high-quality information for land-cover mapping, which in turn enables better management of resources and development of policies that may lead to full development at all levels – e.g. regional, continental, and global. Further, a time series of land-cover maps can contribute immensely to our ability to determine temporal land use dynamics and to minimize the negative impacts that economic development brings to the area and to its available natural resources. For this study, a total of five multispectral images from Landsat TM and

**Table 4.1** Particulars of the study area's topographical sheets

| S. no | Index no. | Survey year | 1st publication | 2nd publication |
|---|---|---|---|---|
| 1 | 34-J/6 | 1954 | 1960 | – |
| 2 | 34-J/7 | 1954 | 1960 | – |
| 3 | 34-J/9 | 1958 | 1960 | – |
| 4 | 34-J/10 | 1954 | 1960 | – |
| 5 | 34-J/11 | 1959 | 1960 | – |
| 6 | 34-J/12 | 1954 | 1960 | – |
| 7 | 34-J/13 | 1954 | 1960 | – |
| 8 | 34-J/14 | 1958 | 1980 | – |
| 9 | 34-J/15 | 1959 | 1960 | 1978 |
| 10 | 34-N/1 | 1954 | 1960 | – |
| 11 | 34-N/2 | 1958 | 1960 | – |

Source: SOP

**Table 4.2** Acquisition history of the utilized Landsat images

| 19 Oct 1989 | 9 Oct 1991 | Sep 1996 | 9 Oct 2000 | Sep 2005 |
|---|---|---|---|---|

Source: Landsat

Landsat ETM (30 m spatial resolution each) for the years 1989–2005 (Table 4.2) were acquired from SUPARCO, USGS, and some secondary users. The orbital path and row of the images was 153 and 39, respectively.

In selection of the Landsat images, three considerations played a role:

(a) Because satellite data are expensive, we depended on those scenes available with SUPARCO, USGS, and some other organizations (GSP, Quetta; UNDP, Quetta) which, although they do not have satellite databanks, luckily possessed the scenes of our AoI for use in their own ongoing projects.
(b) Transparency of the sky during recording of the images by the satellite was also a critical concern, as cloudiness limits identification of land cover.
(c) The cropping calendar of the area has two growing seasons:

- Summer crops (kharif season) – April 15 to November 15 (although vegetative growth stops in November due to cold, enough foliage exists until then on these orchards, which is differentiable on satellite scenes).
- Winter crops (rabi season) – November 15 to April 15. In terms of green foliage, the kharif season is comparatively more lush. Hence, we based our green cover variation analysis only on the images captured in the kharif season.

### Landsat Data: History and Qualities

Landsat-1 was the world's first earth observation satellite (EOS) launched by the USA in 1972. Its excellent set of capabilities emphasized the importance of state-of-the-art remote sensing. Following Landsat-1, Landsat-2, 3, 4, 5, and 7 were launched. Landsat-5 was the first to be equipped with an MSS and a TM. The MSS is an optical sensor designed to observe solar radiation, which is reflected from the earth's surface in four different spectral bands, using a combination of the optical system and the sensor. TM is a more advanced version of the observation equipment used in the MSS; it observes the earth's surface in seven spectral bands, ranging from visible to thermal infrared regions.

The TM is an advanced, multispectral scanning, earth resources sensor designed to achieve higher image resolution, sharper spectral separation, improved geometric fidelity, and greater radiometric accuracy and resolution than the former MSS sensor. This sensor also images a swath that is 185 km (115 miles) wide, but each pixel in a TM scene represents a 30 m × 30 m ground area, except in the case of the far-infrared band 7, which uses a larger (120 m × 120 m) pixel. The TM sensor has seven bands that simultaneously record reflected or emitted radiation from the earth's surface in the blue-green (band 1), green (band 2), red (band 3), near-infrared (band 4), mid-infrared (bands 5 and 7), and the far-infrared (band 6) portions of the electromagnetic spectrum. TM band 2 can detect green reflectance from healthy vegetation, and band 3 is designed for detecting chlorophyll absorption in vegetation. TM band 4 is ideal for near-infrared reflectance peaks in healthy green vegetation, and for detecting water-land interfaces. TM band 1 can penetrate water for bathymetric (water depth) mapping along coastal areas, and is useful for soil–vegetation differentiation, as well as for distinguishing forest types. The two mid-infrared bands on TM are useful for vegetation and soil moisture studies, and for discriminating between rock and mineral types. The far-infrared band on TM is designed to assist in thermal mapping, and for soil moisture and vegetation studies.

Landsat-7, launched on 15 April 1999, is equipped with Enhanced Thematic Mapper Plus (ETM+), the successor of TM. The observation bands are essentially the same seven bands as in the TM, with the addition of the panchromatic band 8, with a high resolution of 15 m. An instrument (Scan Line Converter) failure occurred on 31 May 2003, with the result that all Landsat-7 scenes acquired since 14 July 2003 were collected in 'SLC-off' mode. In that mode, there is a black line (i.e., no data line) on both sides of a scene. These lines are filled in with the previous data set (acquired before 14 July 2003) or with overlap data. This means that the center of the scene is new imagery, but the left and right sides contain old data.

Landsat data have been used by government, commercial, industrial, civilian, and educational communities throughout the world. The data are used to support a wide range of applications in areas such as global change research, agriculture, forestry, geology, resource management, geography, mapping, hydrology, and oceanography. The images can be used to map anthropogenic and natural changes on the earth over periods of several months to two decades. The types of changes that can be identified include agricultural development, deforestation, desertification, natural disasters, urbanization, and the development and degradation of water resources.

Satellite Data Analysis

ERDAS Imagine 9.1 – the Digital Image Processing (DIP) software – was used for interpretation and processing of the images. All thematic maps were developed in ArcGIS 9.1®. Microsoft® Word and Excel were used for documentation and graphical analysis. Imagery analysis follows three basic steps: pre-processing, classification, and output.

*Pre-processing*

The images used were in Transverse Mercator Projection System with Spheroid 1909. They were re-projected into Universal Transverse Mercator (UTM) coordinate system, Zone 42 with Spheroid and Datum as WGS 84 so that the vector files of the target area could be displayed on the satellite images. The images were acquired in EOSAT FAST format. For easy handling and processing, they were imported in ERDAS Imagine native image format (img). In order to reduce processing time and computer resources, it was recommended to subset the satellite image at a desired AoI. Satellite images were truncated at the boundary of the study area. It also provided a high level of confidence in the selection of training sites, hence enhanced the overall accuracy of the results.

Due to inherent low contrast, satellite data require enhancement using various image-enhancement algorithms. Keeping in view the subjective land cover, histogram equalization and standard deviation stretch were applied for the extraction of meaningful information regarding different land cover classes. These algorithms enhanced the low contrast of satellite images and made them more interpretable for further processing. Moreover, the brightness and contrast utility, as well as different false color composites (FCCs) were used to enhance image interpretability.

*Classification*

A satellite image can be defined as a sequential combination of pixels (rows and columns) with stored reflectance values of the ground features at a specific time. Classification of multispectral images is the process of assigning pixels of a continuous raster image to predefined classes (Sabins 1997) or categories of data, based on their data file values. If a pixel satisfies a certain set of criteria, that pixel is assigned to the class that corresponds to those criteria. For land cover mapping, two basic techniques are being used by GIS professionals to classify satellite imageries:

1. *Unsupervised Classification*: This depends upon the data themselves for the definition of classes and is usually used when little is known about the data before classification.
2. *Supervised Classification*: the study at hand has used this technique. It works under the close control of the analyst and involves four basic steps: training samples, feature space examination, signature file production, and output.

(a) **Training Samples**

The first step in the supervised classification of images is 'training'. This is the process of defining the criteria by which patterns in the data are recognized. A training sample is a set of pixels selected to represent a potential class. The data file values for these pixels are used to generate a parametric signature. The computer system is first trained to recognize patterns in the data. In this process, the pixels that represent patterns or land cover features are selected that have been identified using other sources, such as aerial photos, ground truth data, or maps. A training field or training site is the geographical AoI in the image, which is represented by the pixels. The computer system is then instructed to identify all pixels with similar characteristics.

Training samples can be identified using a vector layer that is defining a polygon in the image; or identifying a training sample of contiguous pixels with similar spectral characteristics; or using a class from a thematic raster layer from an image file of the same area (i.e., the result of an unsupervised classification). Through training samples, we identified the areas of high confidence land cover classes on the satellite images and developed thematic maps. These samples were selected by defining criteria for pure class identification on the basis of the results obtained from unsupervised classification, reflectance values, and visual image interpretation.

**(b) Feature Space Examination**

A feature space image, also called a scatter plot, is simply a graph of the data file values of one band of data plotted against the values of another band of data. In ERDAS IMAGINE, a feature space image has the same data structure as a raster image. Each training sample was examined in feature space to ensure its class feasibility.

**(c) Signature File Production**

A signature is a set of data that defines a training sample, a feature space object, or a cluster. A single feature space image, but multiple AoIs, can be used to define the signature. The pixels in the image that correspond to the data file values in the signature (i.e. feature space object) are assigned to that class. The signature files were produced and evaluated for all images and then supervised classification was run over them.

**(d) Output: Thematic Layers Generation**

Temporal thematic maps were generated in ArcGIS 9.1 using the classified images of five different years (Table 4.2). The output land cover maps comprise the following four classes:

- Agriculture – through our personal knowledge of the area, we knew that, in the months corresponding to the images (September, October), almost all the green land cover was tubewell-irrigated agriculture.
- Barren land with salt color – the cultivable-waste covered with a white salt layer washed down from the surrounding mountains and laid.
- Wetland – includes natural streams, irrigation tanks and courses, and the wet surfaces temporarily produced by rainwater.

- Bare soil/settlement/hard rock – includes the cultivable-waste that is free of salt cover, the built-up area as it, being mostly built of mud, marks soil like a spectral signature, and the hard rock protuberances, which are found at the valley's piedmont fringe.

The area of each class was calculated, using ERDAS Imagine Raster Attribute Editor.

### 4.3.2 *Primary Data*

In accordance with the principles of a case-study approach, we collected data by first conducting interactive focus group discussions (FGDs), and next by full-scale structured interviews. We held two FGDs, each containing around ten members. Selection of the members was based on their ability to provide insight into the critical issues related to the problem under study. The participants came from among the local farmers, land owners, and the experienced officials serving in the local water and agriculture related public sector departments. Their open narrative input helped in the subsequent surveys and structuring of the interview protocols.

The study area is characterized by a thin and scattered human population. The number of villages (cadastral units) is large, but on one hand they generally carry too small a population, and on the other hand, their cadastral record is also never properly maintained. Moreover, the magnitudes of spatial variability in the variables under study were small. That is, the variations in agricultural land use and aquifer characteristics were too small to be detected significantly over small distances. For these reasons, we initially intended to adopt the UCs as data units in the whole valley. However, we could do so only in the case of the Pishin district part of the Pishin Valley. In the case of Q. Abdulla District, we were compelled to go for the PCs as data units, instead of the UCs (explained in the text below). A UC (also called a village council) in Pakistan is a grassroot-level political unit (Fig. 4.3) that commonly comprises four to five villages, depending on the size of the population. Normally, a UC represents 10,000–12,000 people. The jurisdiction of a UC is usually part of a tehsil, which itself is a district subdivision.

When we overlapped the map of Pishin Valley over the UC-wise map of the Pishin District, it encompassed 14 of the total 22 rural UCs of the district, plus the Pishin City area, which is administered by a municipal committee. This means that there are a total of 15 (14 + 1 = 15) units of the Pishin district local Government falling within the Pishin Valley boundary. All were examined for land cover variation through the satellite images with temporal intervals of about 5 years. However, the Pishin municipality and the Karbala UC were exempted from the questionnaire administration; the former because it was a non-agricultural urban locality, and the latter because it was mostly and historically barren (Map 5.1).

For Q. Abdullah District, we could not find a UC-wise map from the secondary sources. The available maps were giving boundaries down to the level of PCs only, which are the grassroot-level land revenue areas according to the country's land administration system (Fig. 4.4); the head officer of the local land administration and

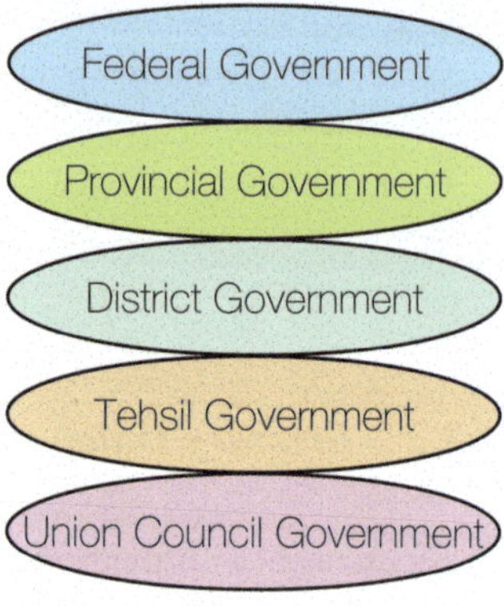


**Fig 4.3** Administrative hierarchy in Pakistan

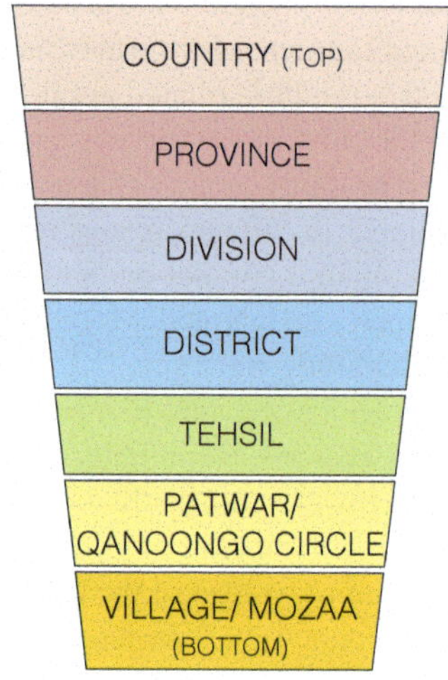


**Fig. 4.4** Hierarchy of land revenue administration in Pakistan

revenue collection is called the 'patwari'. In areas of sporadic agriculture and thin population, like the Q. Abdullah district part of our study area, a PC has nearly the same aerial dimensions as a UC now has per the 2001 Devolution Plan of the Government of Pakistan. For instance, the whole area of the Ajram Shadezai PC constitutes almost the whole of Ajram Shadezai UC. Nevertheless, a PC is commonly slightly larger in area than a UC. For instance, Q. Abdullah district has ten PCs against a total of 15 UCs, which means one PC equals the area of 1.5 UCs. By overlapping the Pishin Valley map over the Q. Abdullah District map, five of its total ten PCs fell inside the Valley's frontiers. Land cover changes of all these five PCs were assessed through the satellite scenes. However, the Ajram Shadezai PC was excluded from the questionnaire survey (Table 4.3) due to its complete and permanent barren character throughout our study period (Map 5.1).

Our primary data sources were two groups of people who were interviewed through two separate structured questionnaire surveys. The groups were as follows:

1. Local farmers/land owners; and
2. Experienced officials (called 'key informants' herein) of the public sector departments concerned.

### Survey A: Farming Group Sampling and Questionnaire Administration

The first survey, which took into account the farmers/landowners, proceeded from April to September 2008. Two complementary approaches were used:

1. First, two interactive FGDs were arranged (one each in Pishin and Q. Abdullah districts) to understand the perceptions of the masses about the major issues affecting the agricultural and groundwater economy of the area. For the purpose of the dialogues, a checklist of anticipated issues was prepared that was invoked issue by issue on the table to elicit participants' views. The interactive groups comprised the farmers/landowners and the locally deputed expert officials. By and large, the methodology suggested by Ahmad and Ahmad (2007) and Ahmad (Ahmad 2007a, b) was applied, as it has been tried in various districts of Balochistan by the ADB-sponsored consultant-based studies. Through these dialogues, consensus information was deduced, on the basis of which specific areas for further inquiry were identified and incorporated in the two different interview schedules for eliciting formal and statistically representative data.
2. Second, a farm-level survey was conducted. We initially had intended to survey every UC/PC falling within Pishin Valley's frontiers (i.e. a census, not sampling); however, when satellite scenes showed a few of them to be overwhelmingly non-agricultural, we then had to drop them from the questionnaire administration. Given this, we disregarded two of the 15 political units – UCs – in the Pishin district part of Pishin Valley, but only one of the five land revenue units – PC – in Q. Abdullah district part of the Valley.

During the timeframe of our study (1981–2008), there were three types of farmers in the surveyed UCs/PCs: (1) those with all farms irrigated; (2) those with all farms rain-fed; and (3) those with some farms irrigated and some rain-fed. Our target population included only the first and third categories of the farmers. The second category – all farms rainfed – was not included because they had no role in the exploitation of groundwater for irrigation, and were not ever affected by its deterioration in agricultural matters. The official cadastral record of the study region had several limitations. It was neither complete nor reliable and, above all, the concerned agencies expressed reluctance to give access to it, probably because they feared exposure of their inefficiency and misconduct. Therefore, we ourselves estimated the target population – all tubewell irrigation farmers – in all target UCs/PCs through narration of the local farmers in their respective component villages. As an exception, where some of the villages were either so small or far-flung and hardly accessible, their target populations were roughly estimated through reports from farmers who were

**Table 4.3** Location of the sampled villages in Pishin Valley

| S. no. | District | UC/PC | Area (km$^2$) | Sampled villages | Location of sampled villages Latitude (N) | Longitude (E) |
|---|---|---|---|---|---|---|
| 1 | Pishin | Alizai | 34.8 | Alizai | 30°43′58.235″ | 66°57′3.632″ |
| 2 | – | Batezai | 67.1 | Bianzai | 30°37′19.259″ | 67°3′31.957″ |
| 3 | – | D. Khanzai | 32.9 | Mian Khanzai | 30°43′50.577″ | 67°5′23.794″ |
| 4 | – | Gangalzai | 21.1 | Shinghari | 30°42′32.55″ | 66°51′47.005″ |
| 5 | – | Huramzai | 44.7 | Huramzai | 30°43′31.885″ | 66°50′23.001″ |
| 6 | – | Malakyar | 51.3 | Kamalzai | 30°41′20.748″ | 67°7′35.141″ |
| 7 | – | Manzaki | 49 | Sarak Umarzai | 30°44′40.386″ | 67°8′43.218″ |
| 8 | – | Manzari | 57.7 | Salad Nika | 30°44′50.991″ | 66°59′45.464″ |
| 9 | – | N. Malezai | 110.4 | Malezai | 30°40′39.409″ | 66°58′11.372″ |
| 10 | – | Saranan | 94.7 | Haikalzai | 30°36′37.945″ | 66°54′59.2″ |
| 11 | – | Simzai | 54 | Shakarzai | 30°44′6.942″ | 66°54′27.217″ |
| 12 | – | Tora Shah | 56.8 | Tora Shah | 30°40′29.392″ | 67°6′2.728″ |
| 13 | – | Yaru | 49.4 | Razzaq Kili, Tora Ghundai | 30°29′59.566″ | 66°56′59.719″ |
| 14 | Q. Abdullah | Gulistan | 233.3 | Habibzai | 30°40′4.962″ | 66°39′38.028″ |
| 15 | – | Maizai | 64.1 | Majak | 30°41′30.764″ | 66°45′5.784″ |
| 16 | – | Q. Abdullah | 134.2 | Lamarhan | 30°42′41.734″ | 66°43′33.665″ |
| 17 | – | Segi | 368.1 | Zakariazai, Mirbaz | 30°27′23.903″ | 66°31′2.76″ |

Source: GIS calculation and field survey

non-local but familiar with the area. Statistics of the villages were aggregated to derive size of the total target population in the individual UCs/PCs, and also in the entire study area (Table 4.4). Given the time and resources available, we decided to sample 6 % from the target population, a total of 2,962 individuals. Thus, through this arbitrary arrangement, our gross sample size concluded at 178.

Next, each UC/PC was considered as a separate population stratum, and the stratified random sampling method was applied. This sampling technique actually means the independent simple random samples taken within each stratum. Because the 17 strata differed in size, we applied the 'formula of proportional allocation of sample sizes' to derive statistically representative sample sizes for each independent stratum (UC/PC). The formula is given hereunder:

$$n_h = \left(N_h / N\right) n$$

Where:

$n_h$ = required sample size in the $h$th UC/PC.
$N_h$ = size of the sampling frame (total number of irrigation farmers in this case) in the $h$th UC/PC.
$N$ = total number of sample units in the target population.
$n$ = total number of desired sample units in the target population (total number of irrigation farmers in the 17 UCs/PCs in this case).

**Table 4.4** Proportional allocation schemes for stratified random sampling of $n = 178$ elements from an $N = 2,962$ element population

| S. no. | Stratum $h$ (*UC/PC*) | Stratum size $N_h$ | Proportional allocation | |
|---|---|---|---|---|
| | | | $N_h/N$ | Sample size $n_h$ |
| 1 | Alizai | 149 | 0.05 | 9 |
| 2 | Batezai | 138 | 0.047 | 8 |
| 3 | D. Khanzai | 201 | 0.068 | 12 |
| 4 | Gangalzai | 182 | 0.061 | 11 |
| 5 | Huramzai | 197 | 0.067 | 12 |
| 6 | Malakyar | 187 | 0.063 | 11 |
| 7 | Manzaki | 103 | 0.035 | 6 |
| 8 | Manzari | 119 | 0.04 | 7 |
| 9 | N. Malezai | 160 | 0.054 | 10 |
| 10 | Saranan | 118 | 0.04 | 7 |
| 11 | Simzai | 133 | 0.045 | 8 |
| 12 | Tora Shah | 258 | 0.087 | 16 |
| 13 | Yaro | 147 | 0.05 | 9 |
| 14 | Gulistan | 256 | 0.086 | 15 |
| 15 | Maizai | 182 | 0.061 | 11 |
| 16 | Q. Abdullah | 124 | 0.042 | 7 |
| 17 | Segi | 308 | 0.1 | 19 |

Source: Saeed (2012)

This technique of sampling is also referred to as 'proportionate stratified sampling'. Proportional allocation is, in fact, the simplest allocation procedure and is widely used.

Since the nature of our study was such that the required attributes in the population were virtually non-variant over short distances (village to village), it was therefore logical to select any one or two villages in every UC/PC and draw the whole allocated sample size from there to represent the entire relative UC/PC. The justification of such a practice is based on two main arguments:

- It could serve well our cartographic requirement, since the isopleth map-based analysis (e.g. water depth contouring) demanded some grid-wise defined points inside the broad UC/ PC polygons.
- It was the only feasible option as explained below.

The villages were selected via purposive sampling method (Map 4.1). This method, occasionally referred to as 'judgment sampling', is a sampling design wherein the sampling units are selected subjectively by the researcher if they appear to be representative of the whole population. The following judgments provided the basis for selection of the sampled villages:

- The selected village, if not the largest, should be at least one of the largest in the relative UC/PC.
- The village should be located in that part of the relative UC/PC which has a long history (at least of 28 years) of electrification and a widespread land use of tubewell-irrigated agriculture.

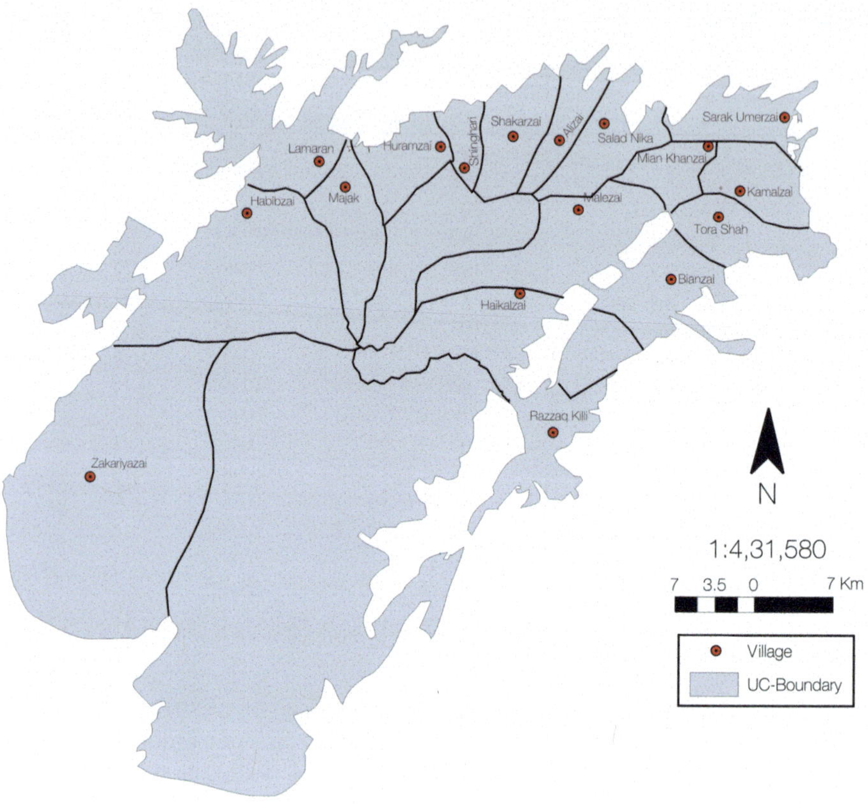


**Map 4.1** Location of the sampled villages in their respective UCs/PCs

- Personal rapport with the local facilitator was also desirable, because, without this, genuine cooperation from all the frame farmers in any selected village was doubtful. In the case of fragmented participation of farmers, we could not have truly obeyed the predesigned procedure of simple random sampling.

Supported by the above logic, we enlisted the names of all the tubewell irrigation farmers in the selected villages. In the enlisting process in each village, the source of information was a well known farmer in the village. In the case of large villages, e.g. Segi and Torashah, the lists were also cross-checked through some other locals. Next, we picked the respondents using the simple random method (lucky draw of their names). Simple random sampling is considered the basic form of probability sampling in which each element of the population frame has an equal probability of selection.

The survey tool was a structured questionnaire administered via personal (face-to-face) interview. The tool was designed to elicit time-series information on socioeconomic variables (age, family size, farmers' education, non-agricultural occupations and financial share of the farming in the household's livelihood); land use variables (farm size, cultivated land, cultivable-waste, irrigated/un-irrigated land), etc. Further, the questionnaire asked about some irrigation variables (e.g., watertable fall, horsepower of the prime mover [electric motor], water quality indicators, irrigation cost, structure of water-conveyance channels, micro-irrigation technology, farm-income, and changes in crop selection). Toward the end of the questionnaire, the farmers were provided with open-ended questions to enable them to express their views about the merits and demerits of the subsidy that is being given in electricity tariff to irrigation tubewells. Finally, they were asked to give suggestions as to how the available groundwater resources could be conserved and enhanced (Appendix F). The questionnaire was pre-tested using the method suggested by Reynold and Dimantopoulos (1998) in order to detect and remove any deficiencies therein, and thus render it appropriate for full-scale survey.

The total 28-year timeframe of the study (1981–2008) was sliced into four eras with increasing lengths toward the past (first and second eras of 4 years each; third and fourth eras of 10 years each). We chose this methodology because we found that, the further back in time we asked participants to remember, the more difficult it became for them to recall the precise timing of any change. In fact, in the pre-testing phase of the questionnaire, we realized that a great number of the respondents were embarrassed when they were asked to report any change in the more distant past when we used short time ranges (e.g., only 4 years). For cartographic presentation, results of the analysis of the sampled villages were generalized to represent their whole respective UCs/PCs.

### Survey B: Line Expert Interviews

The survey of line experts was carried out from December 2008 to March 2009. A three-page questionnaire was distributed, mostly personally, but occasionally by mail, to the expert officials from water resource and agriculture departments posted at the Quetta headquarters, as well as at sub-offices in the Pishin and Q. Abdullah districts. Even if questionnaires were personally delivered, responses were not collected face-to-face. Rather, participants were allowed several days to sit in private and feedback their genuine knowledge and expertise. Most questionnaires were collected back after a week or more. The need for and utility of the study, as well as some tricky points within the questionnaire, were explained to the participants through a covering letter.

Of a total 26 distributed questionnaires, only 15 were completed and returned by post, giving a response ratio of 58 %. The questionnaire contained 15 closed-ended questions with multiple possible answers. The multiple choices to each and every question were measured on a Likert ordinal scale. Toward the end of the questionnaire, some open-ended queries were also used to draw upon the experts' free thinking on the critical issues and their corrective suggestions in this respect (Appendix G).

### Primary Data Analysis

Each of the two primary data sets (farmers and experts) was first tabulated in a Microsoft® Excel worksheet. Frequency tables, averages, percentages, and various types of graphs were also prepared in Microsoft® Excel.

To assess the economic sustainability of tubewell-irrigated agriculture, the straight-line method of budgetary analysis, as suggested by Adegeye and Dittoh (1985), was performed. On one hand, this method is easy to compute statistically, and, on the other hand, is safer to execute socially. This is because this method does not ask the field managers in direct terms for their net profit, rather, it is computed indirectly through a deceiving course – expenditure list and gross values of products. The formula per that method is given hereunder:

$$\text{Total Cost }(\text{TC}) = \text{Total Variable Cost }(\text{TVC}) + \text{Total Fixed Cost}$$

$$\text{Net Profit }(\text{Net Income}) = \text{Total Revenue }(\text{TR}) - \text{Total Cost }(\text{TC})$$

## 4.4 Summary

Observational methods of data collection are suitable for investigating those phenomena that can be observed directly by the researcher. However, not all phenomena are accessible to direct observation by an investigator; therefore, researchers often need to collect data by asking people with direct knowledge of a phenomenon to reconstruct it for others. The researcher approaches and interviews samples of individuals; the obtained responses constitute the data upon which hypotheses are examined. This is the methodology of a questionnaire survey research; fundamentally, the same was exercised in our study and presented accordingly in this chapter. In addition, to reinforce the reliability of the questionnaire data, the observational method is also employed through 'eye in the sky' – satellite images. This chapter introduced the data types and sources; the remotely sensed data in particular was discussed at much length. It explained the status of UCs and PCs in the administrative structure of the country (our sampling units), and elaborated the complete timeline and procedure of how we collected and analysed the archival and primary data.

## References

Adegeye AJ, Dittoh JS (1985) Essential of agricultural economics. Impact Publishers, Nig. Ltd, Ibadan, pp 63–67

Ahmad S (2007a) New vision and strategy for managing water and energy use in tubewell irrigated agriculture of Balochistan. TA-4560 (Pak), water for Balochistan policy briefings, vol 3. Govt. of Balochistan, ADB, and Royal Netherlands Govt, Quetta

Ahmad, S (2007b) Why diagnostic process is essential for designing research for development – improving performance of irrigated agriculture in Pakistan. TA-4560 (Pak), vol. 1, No. 3, Quetta

Ahmad I, Ahmad S (2007) Water productivity and economic efficiency of tubewell irrigated farms in Balochistan – issues and policy reforms. TA-4560 (Pak), water for Balochistan policy briefings, vol 3. Govt. of Balochistan, ADB, and Royal Netherlands Govt, Quetta

Sabins FF Jr (1997) Remote sensing: principles and interpretation, 3rd edn. W. H. Freeman & Co, New York

Saeed A (2012) Inter-effects of tubewell irrigated agriculture and the aquifer potential in Balochistan: a case study of Pishin Valley, 1981–2008. Ph.D. thesis, University of the Punjab, Lahore, Pakistan

# Chapter 5
# Results and Discussion: Part A

**Abstract** The widespread rural electrification during the 1970s and 1980s attracted tubewell technology, which boosted the area's agricultural economy. However, groundwater pumping soon became over-pumping, causing deterioration of the aquifer. Since groundwater has historically been the central source of irrigation here, its deterioration affected agriculture in several ways. This chapter examines land use dynamics of tubewell agriculture, first through field survey data results of which are then validated by satellite images. The phrase 'agricultural development' means growth (new additions) and modification (re-alignment) of an agricultural system. Therefore, this chapter incorporates horizontal (expansion) and vertical (intensification) growth trends in agriculture, cropping pattern alterations, and the economic performance of farming in the face of high irrigation costs. The spreading desertification process has been highlighted both by the field survey results and by satellite image classification. Since farmers are the key stakeholders in the agricultural enterprise, their socio-economic features are unveiled first of all.

**Keywords** Pishin Valley • Results • Households • Respondents • Imageries

## 5.1 Socio-economic Features of Respondents

In the 178 sampled households of 17 UCs/PCs, the gross average household size is 32, which indicates large family units and the joint family system where married brothers (even cousins in some cases) live combined and share combined earnings (Table 5.1). Further, an average household contributes 3.6 male adults to fulltime farming, who handle an average of 17.9 acres of irrigated land per head. A total of 32 % of the farmers are illiterate. Among the 68 % literate, 18 % have education up to primary level (1–5 years of schooling), 41.5 % up to high school level (6–10 years), and only 8 % are college/university graduates.

A.S. Khattak, *Mutual Sustainability of Tubewell Farming and Aquifers: Perspectives from Balochistan, Pakistan*, Advances in Asian Human-Environmental Research, DOI 10.1007/978-3-319-02804-0_5, 

**Table 5.1** Socio-economic profile of the respondents (farmers), 2008

| UC/PC | Household size (avg.) | Farm workers per household (avg.) | Acres per farmer (avg.) | Farmers' literacy levels (%) | | | |
|---|---|---|---|---|---|---|---|
| | | | | Illiterate | Primary school | High school | College/ University |
| Yaro | 23 | 2.3 | 34.6 | 38 | 4.8 | 38 | 19 |
| Batezai | 39.5 | 2.1 | 15 | 35 | 23.5 | 41 | – |
| Manzari | 54 | 9.3 | 3 | 94 | 1.5 | 1.5 | 3 |
| Malakyar | 43 | 5.9 | 9 | 23 | 27.7 | 47.7 | 1.5 |
| Torashah | 27.6 | 3.5 | 8 | 16 | 9 | 64 | 10.7 |
| Saranan | 26 | 2.4 | 34 | 23.5 | – | 53 | 23.5 |
| Manzaki | 41.3 | 1.5 | 30.5 | – | – | 66.7 | 33 |
| D. Khanzai | 43.3 | 4.7 | 7 | 18 | 32 | 37.5 | 12.5 |
| N. Malezai | 20 | 2 | 38 | 42 | – | 47.4 | 10.5 |
| Huramzai | 34 | 2 | 27 | 71.4 | 9.5 | 19 | – |
| Gangalzai | 35.4 | 5.1 | 6 | 16 | 55.4 | 27 | 2 |
| Simzai | 25.6 | 3 | 19 | – | 34.8 | 65 | – |
| Alizai | 35 | 3.8 | 9 | 8.8 | 35 | 50 | 6 |
| Gulistan | 27 | 4.2 | 5.6 | 46 | 16 | 35 | 3 |
| Q. Abdullah | 27.3 | 3.2 | 7.4 | 68.4 | – | 21 | 10.5 |
| Maizai | 21.6 | 3.8 | 7.4 | 31.8 | 18 | 47.7 | 2.3 |
| Segi | 23 | 2.3 | 42.4 | 11.6 | 42 | 44 | 2.3 |
| **Valley Avg.** | **32** | **3.6** | **18** | **32** | **18** | **41.5** | **8** |

Source: Field Survey

The prevailing common-income lifestyle, coupled with low literacy, may be among the reasons for adherence to the traditional cropping and irrigation methods in the area. Since the farmers, due to lack of education and awareness, have little consciousness of the long-term sustainability of their farm-based income, they do not care to voluntarily adopt water-efficient modern irrigation technologies. In fact, experts do not recommend that the area continue with agriculture as it is, by its very nature, a highly water-intensive business and is likely to cause pot-water crisis in the not very distant future. However, by and large, locals have no alternative income sources because the availability of appropriate non-farming jobs appears low due to their poor literacy and given competitiveness in the modern sophisticated world.

Due to population growth, dwindled man-cultivated land ratio and farm profitability, financial dependence on farms has been gradually decreasing over time. Table 5.2 indicates that during the 1989–1990 decade, there were 46.6 % households whose sole (or nearly so) bread winner was their farm. This ratio dropped gradually in every following data period, finally reaching 37 % in 2006–08. Table 5.2 shows that, during 1981–90, when deep pumping technology (submersible motors) was not yet widely adopted, there was a relatively greater percentage of people lost their farm (6.2 %) due to aquifer drawdown beyond the capacity of the traditional irrigation means – dugwells, karezes, and non-submersible (surface-mounted) tubewell pumps. The data since 2000 show gradual revision of the desertification process, reflecting the effects of the 1998–2004 drought and depletion of the aquifer, either in terms of water's extinction, brackishness, or dropping too deep to be tapped within a positive cost-benefit range.

**Table 5.2** Percent frequency of households by farm income share in their annual budget

| | Farm share | | | | | | | | | | |
|---|---|---|---|---|---|---|---|---|---|---|---|
| Periods | Farm deserted | 1–10 | 11–20 | 21–30 | 31–40 | 41–50 | 51–60 | 61–70 | 71–80 | 81–90 | 91–100 |
| 1981–90 | 6.2 | 8.4 | 10.7 | – | 5 | 12.4 | 1.7 | – | 9 | – | 46.6 |
| 1991–2000 | 0.6 | 5 | 8.4 | 7.9 | – | 14.6 | 0.6 | 4.5 | 10 | 4 | 44.4 |
| 2001–04 | 1.7 | 9.5 | 11 | 10 | 9 | 8 | – | 0.6 | 8.4 | 3 | 39 |
| 2005–08 | 5.6 | 10.7 | 11 | 14 | 6 | 4 | – | 4.5 | 4 | 3 | 37 |

Source: Field Survey

**Table 5.3** Occupational structure of the farming households, 2008

| | | | Complementary non-farm occupations (% of column 3) | | | | |
|---|---|---|---|---|---|---|---|
| UC/PC | Fully farm-dependent households (avg. %) | Partially farm-dependent households (avg. %) | Trade | Govt. services | Transport | Daily waging | Civil works contracting |
| Yaro | 11 | 89 | 37.5 | 12.5 | 12.5 | – | 37.5 |
| Batezai | 12.5 | 87.5 | 71.4 | 14 | 14 | – | – |
| Manzari | 57 | 43 | 66.7 | – | 33.3 | – | – |
| Malakyar | 72.7 | 27.3 | 66.7 | 33.3 | – | – | – |
| Torashah | 43.8 | 56 | 22.2 | 22 | 11 | 44.4 | – |
| Saranan | – | 100 | 43 | 43 | – | – | 14.3 |
| Manzaki | – | 100 | 16.7 | 66.7 | – | 16.7 | – |
| D. Khanzai | 50 | 50 | 66.7 | 16.7 | – | 16.7 | – |
| N. Malezai | – | 100 | 80 | 10 | 10 | – | – |
| Huramzai | 25 | 75 | 77.8 | 11 | – | 11 | – |
| Gangalzai | 54.5 | 45.5 | 60 | – | 40 | – | – |
| Simzai | 25 | 75 | 100 | – | – | – | – |
| Alizai | 44.4 | 55.6 | 60 | – | – | 40 | – |
| Gulistan | 46.7 | 53.3 | – | 12.5 | 62.5 | 25 | – |
| Q. Abdullah | – | 100 | 71.4 | 14.3 | – | 14.3 | – |
| Maizai | 45.5 | 54.5 | 50 | 16.7 | 33.3 | – | – |
| Segi | 47.4 | 52.6 | 80 | 10 | 10 | – | – |
| **Gross %** | **31.5** | **68.5** | **57.1** | **16.7** | **13.3** | **9.9** | **3.0** |

(Source: Field Survey)

Due to diminishing compatibility of farm income to needs, dependence on farming as the sole income generator has declined. In 2008, when the field survey was conducted, 31.5 % of households were fully supported by their farm income (Table 5.3). The remaining 68.5 % had part of their workforce employed in other jobs, like trade, government jobs, transport services, daily labor in construction works, etc., and civil works contracting.

Table 5.3 further reveals that full or partial dependency on farm income greatly varies among the different UCs/PCs. For instance, in Malakyar UC of Pishin district, those fully dependent on farm income are the highest, followed by Manzari UC. Both of these areas are located in the north-eastern flank of the valley, close to BKK and other DADs, which have been reported as slightly contributive in aquifer recharge, hence the level of farm income is high enough there. It is meaningful to note that the UCs/PCs, which are farther from the recharge dams such as Yaro, Saranan, and N. Malezai, etc., generally have no/or few households able to sustain a complete living from their farms. Regarding the involvement of farmers in side businesses, it can be hypothesized that when a farming family is also involved in non-farm businesses, its farming performance is affected and productivity is reduced, especially in arid area agriculture like the Pishin Valley.

## 5.2 Aquifer Depletion Affecting Tubewell Farming

### *5.2.1 Physical Growth of Tubewell Farming*

By physical growth of tubewell farming we mean two things: (1) expansion in tubewell irrigated area, and (2) farm and/or crop intensification under tubewell irrigation. The former objective can be achieved by bringing virgin lands directly to tubewell farming and/or extending tubewell irrigation facilities to previously rain-fed lands. The latter objective requires high inputs of irrigation, fertilizer, and labor; farm and/or crop intensification is theoretically feasible wherever these are available.

#### Expansion of Tubewell Farming

For us, this concept means situation analysis of the increase or decrease in land under tubewell irrigation.

Change in Under-Cultivation Farm Size: Survey Data

The areal extent of agricultural land use is estimated through time-series field survey data and archival satellite images. While the latter showed only the existing green cover, the former was aimed at evaluating time series changes in under-cultivation (net-sown plus currently fallow) farm size. Thus, per the survey data, the average under-cultivation farm size was 63.3 % of the total land holding (cultivable-waste and the land not available for cultivation being 36.7 %) per household during 1981–90 (Fig. 5.1). During the next decade of 1991–2000, this ratio increased to 67.6 % at the expense of the cultivable-waste, giving a growth rate of 4.3 % per 10 years, or 0.43 % per annum. During the whole study period of 28 years, this was the only positive trend. Since then, the column of under-cultivation land is gradually reducing and that of the cultivable-waste rising accordingly. This finding is indicative of the process of desertification. Note that the under-irrigation area during 2001–04 shows an increase from its previous size, whereas, in the same period, the under-cultivation area has decreased. The explanation is that the under-cultivation area includes both irrigated and rain-fed farming; hence when as a whole it was

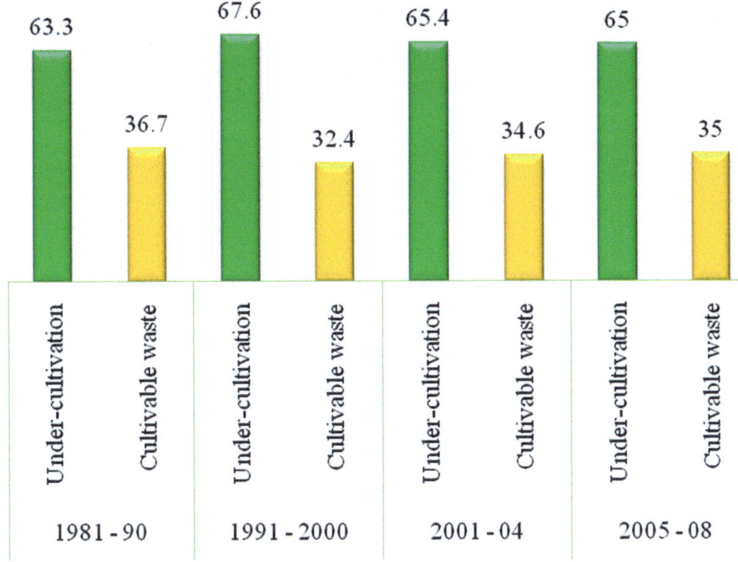


**Fig. 5.1** Percentage ratio of under-cultivation land to cultivable-waste land per household (Source: Field Survey)

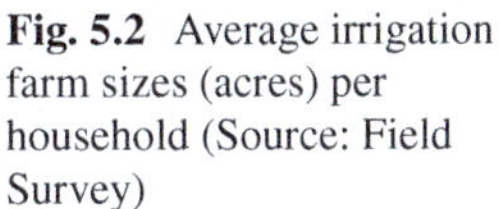

**Fig. 5.2** Average irrigation farm sizes (acres) per household (Source: Field Survey)

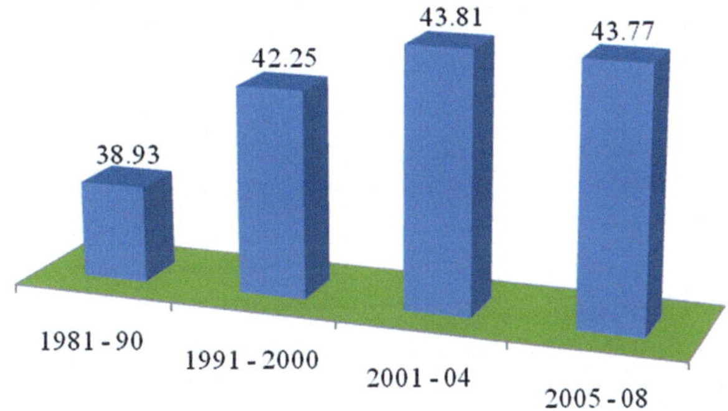


decreasing, some of the rain-fed farms were also being subjected to irrigation because rain-fed farming became economically unproductive due to unreliability of precipitation and high input costs.

Change in Irrigation Farm Size: Survey Data

Average irrigation farm size per household was 38.93 acres during 1981–90 (Fig. 5.2). In the next decade (1991–2000), it rose to 42.25 acres, a rise of 0.33 acres per year. In the next 4 years (2001–04), per year increase in farm size was 0.39 acres

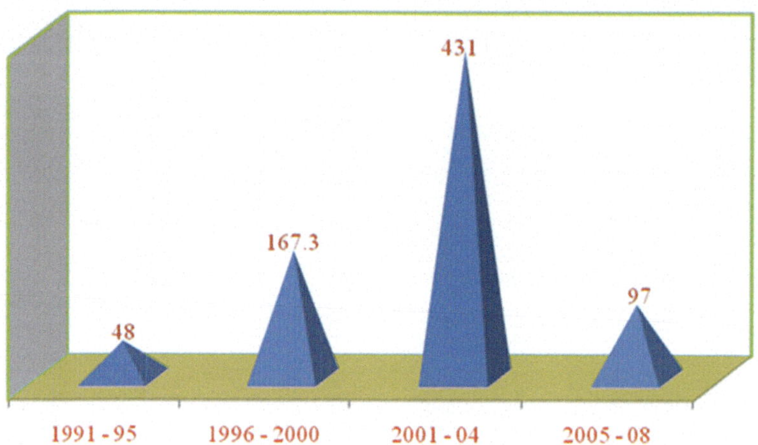


**Fig. 5.3** Total desertified land (acres) in Pishin Valley during 1991–2008 (Source: Field Survey)

per head (household), an accelerated growth rate compared with the previous timeframe. Since the year 2004, a decline is recorded, as between 2005 and 08 the irrigation farm size decreased by a rate of 0.01 acres per head per annum.

From the field knowledge, we know that the increase in the average irrigated farm size until 2004 came about as a result of both bringing new land under plow with irrigation, and extension of irrigation facilities to previously rain-fed lands. The data further derive that tubewell farming expanded by an average of 5 acres per head over the total span of our study period (Fig. 5.2). This may be considered a slow pace of growth in the perspectives of the availability of abundant cultivable land and substantial population rise; however, growth has been consistent (positive tendency) until 2004. The apparent main reasons being the expansion of rural electrification, improving pumping technology, rising market values of the products, and extension of communication and transportation facilities to enter remote markets. The slight negative trend in the farm size from 2004 onwards can be attributed to aquifer deterioration and the consequent desertification phenomenon.

While agriculture as a whole has registered a more or less rising trend during our study period, there has been, at the same time, some unlucky farmers in the sample who lost a part of their previously productive lands to desertification or, to say softly, to the category of cultivable-waste because of water scarcity. We aggregated the amount of land belonging to sample respondents that had been deserted during various periods (Fig. 5.3).

The farmers, to the best of their memory, indicated that the desertification phenomenon emerged around the year 1991, gradually strengthening until 2004. There was a quantum leap (263.7 acres) in the deserted acreage between the years 2000 and 2004 ($471-167.3=263.7$ acres), reflecting the impact of the acute dry spell during that period, coupled with the perpetuating depletion of the watertable and related drying out of tubewells. Nonetheless, from 2004 onwards, during 2005–08, we see a tendency towards stabilization in the amount of the under-cultivation land as the total land abandoned from cultivation during that era was only 97 acres across the entire valley.

Change in Net Sown Area: Satellite Data

The satellite images provide point data of the net sown area in summer cropping season (Maps 5.1 and 5.2). These data corroborate the survey findings to a great extent (Table 5.4). On the 1989 image, we see 27,962 green acres (4.7 % of the total acreage) of the total 597,079 acres area of the valley. This amount almost doubled in only the next 2 years to 1991. This means that 10,413 acres of cultivated lands were added annually during 1989–91. In 1996, we again see an increase in the net sown area but not as quickly as during the former period. During the 5-year duration of 1991–96, the total added was 10,230 acres (2,046 acres per year), which is little addition than, even, that added between the 2-year span of 1989–91. The year 2000 imagery presents a grim picture: between the year 1996 and 2000, the Valley had been subscribing 4,591 acres (59,017 − 40,655 = 18,362 divided by 4 = 4,590.5) of irrigated land annually to cultivable-waste, which may be considered temporary desertification.

For the slower pace of agricultural growth during 1991–96 and its decline during 1996–2000, two reasons may be argued:

1. The years 1996 and 2000, assessed through satellite scenes, are low precipitation years; in particular, the year 2000 is the second driest (only 3.4 in. ppt.) among the 10 years from 1996 to 2005 (Table 3.2).
2. The tubewell data show that when the groundwater irrigation boosted agriculture until the year 1996, the watertable dropped considerably, causing many tubewells to dry out. Consequently, resource-poor farmers could not arrange timely replacements, and thus their farms turned barren.

Table 5.3 shows the year 2005 as having the greatest cultivation: 10.5 % of the valley's total area. This may be due to the following:

- 2005 is the wettest year (11.54 in. ppt.) compared with other years in the Table 3.2. It may have yielded exceptionally high cultivation. Further, it is likely that some natural greenery grew on cultivable-waste lands that was sensed by the satellite and incorporated to the real agricultural green cover statistics.
- Post-2001, once again much foreign aid was directed to the country due to US military action in Afghanistan. Having resources, the Government funded installation of tubewells for irrigation, which drew deep drilling and deep pumping technologies to the area. It enabled farmers to not only recapture some deserted farms, but also to bring new lands under plow.

The landcover of all the 17 surveyed UCs/PCs was also individually assessed from Landsat images (Table 5.5, Map 5.2). It is common in all images that 2000 is

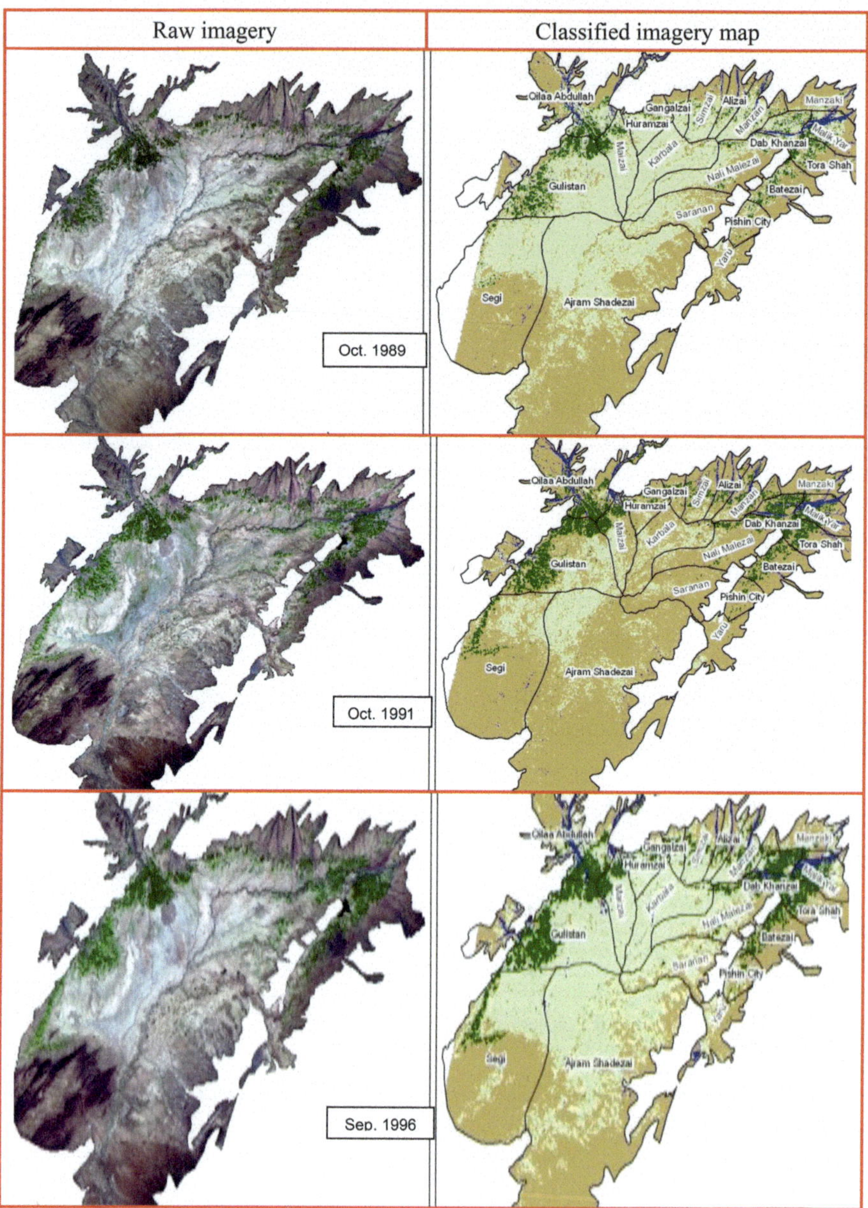


**Map 5.1** Mosaic of Pishin Valley's land cover, Landsat scenes, 1989–2005

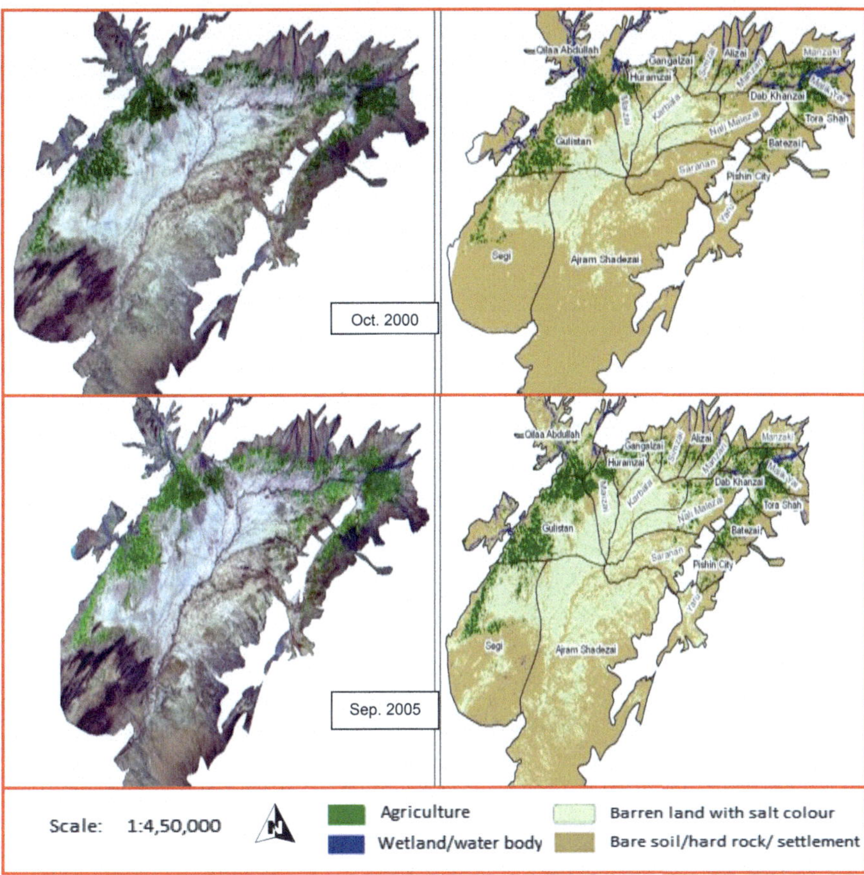


**Map 5.1** (continued)

the least cultivated year. Similarly, all the UCs/PCs show rising trends of cultivation from 1989 onwards to 1991 and 1996. Although the year 2005 dominates 1996 in the amount of cultivated land when viewed for the valley as a whole, there is enough disparity in this regard among the individual UCs/PCs. For instance, Q. Abdullah, D. Khanzai, Alizai, and Torashah have greater areas under cultivation in 1996 than 2005. This means the watertable in these four localities was less depleted by tubewell irrigation after 1991 than were other localities. This may be because these UCs/PCs are located in the northern piedmont flank of the valley – a zone with the following privileges:

- Here there is a concentration of DADs (Map 3.8). In particular, D. Khanzai and Torashah are almost on the brink of the big historical dam: BKK.
- Since the surface water streams enter the valley from this direction, they contribute relatively more to aquifer recharge in this zone.

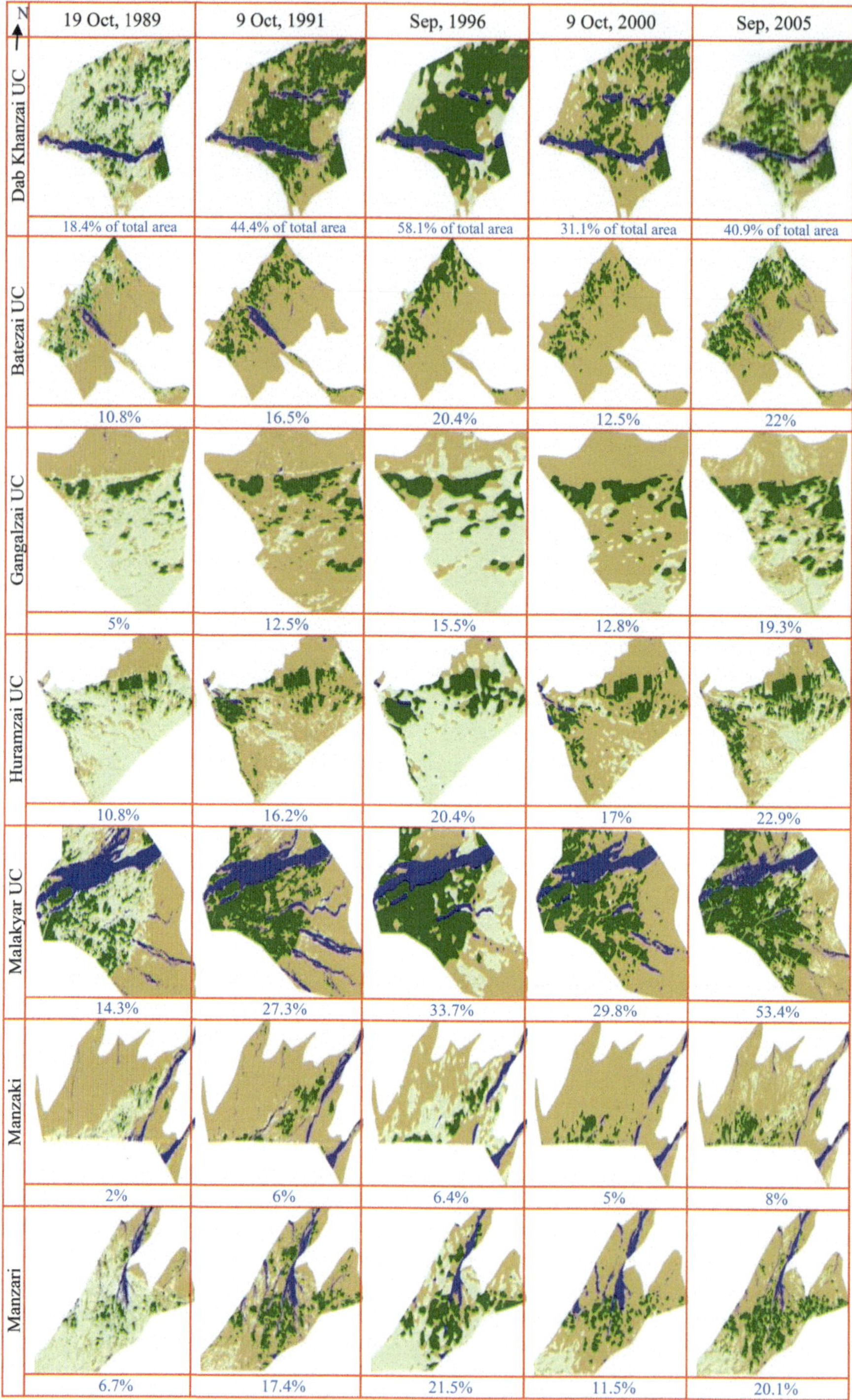


**Map 5.2** Mosaic of UC/PC-wise cultivated land (green), Landsat scenes, 1989–2005

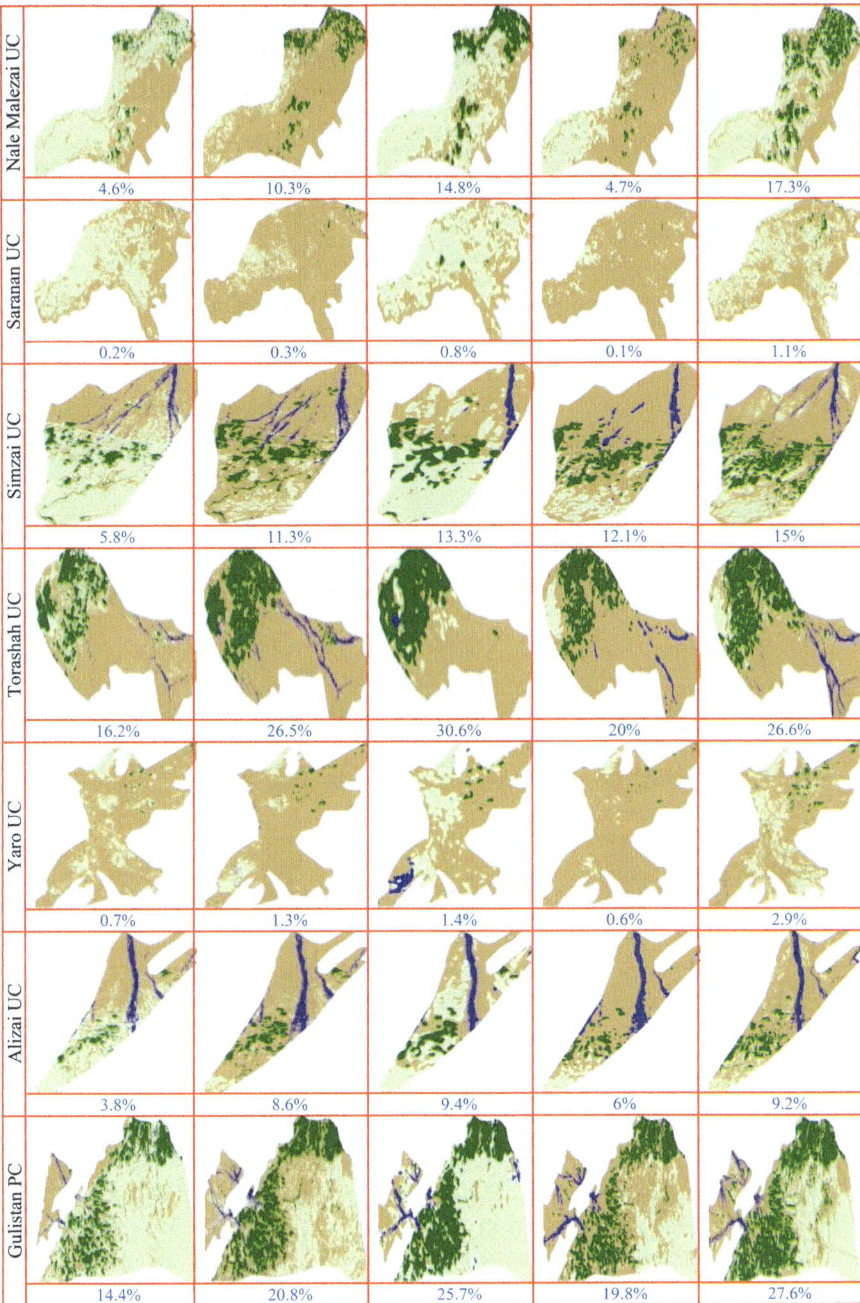


**Map 5.2** (continued)

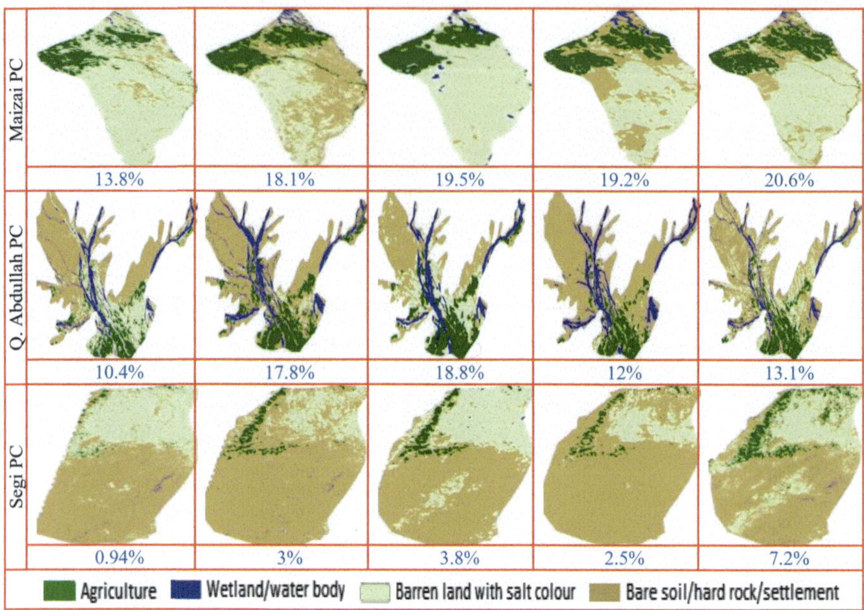


**Map 5.2** (continued)

- Groundwater gradient (flow direction) is from this side down the valley. In this sense, the tubewells here are the upper riparian to those further down the valley, hence have priority opportunities in water abstraction.

## Change in Farm Intensity

Farm-fallowing and intercropping are two main criteria that can be used to evaluate farm intensity. Because of the limitations of credible time-series secondary data, we collected only point data for 2008 on these two parameters.

### Farm Fallowing

The practice of farm fallowing was found to vary in most of the UCs/PCs (Table 5.6). On average, the number of farmers forced to fallow a part of their farms was greater in the kharif season (32 %) than in the rabi season (27.6 %). This means that 68 % of the total farmers in the kharif season and 71.4 % in the rabi season were cultivating with 100 % farm intensity. Since kharif is the summer season, more farm fallowing is understandable given the scarcity of water. However, we did not use satellite images for the rabi season (when farm intensity is greater) because since most of the cultivation in the study area is horticulture, the type of images available to us could not identify due to the lack of green foliage in this season. Further,

**Table 5.4** Area (acres) under various land cover/use categories (valley aggregate)

| Land use categories | 1989 | | 1991 | | 1996 | | 2000 | | 2005 | | Rise/fall, 1989–2005 |
|---|---|---|---|---|---|---|---|---|---|---|---|
| | Acres | % of total | Acres | % of total | Acres | % of total | Acres | % of total | Acres | % of total | |
| Cultivated land | 27,962 | 4.7 | 48,787 | 8.2 | 59,017 | 9.9 | 40,655 | 6.8 | 62,467 | 10.5 | 34,505 |
| Salt colored barren land | 232,118 | 38.9 | 81,844 | 13.7 | 249,991 | 41.9 | 88,645 | 14.8 | 178,540 | 29.9 | –53,578 |
| Bare soil/settlement/hard rocks | 304,924 | – | 449,408 | – | 273,641 | – | 453,715 | – | 349,021 | – | 44,097 |
| Wetlands | 8,443 | – | 11,890 | – | 11,292 | – | 10,766 | – | 8,283 | – | –160 |

Source: Statistics derived from the Landsat scenes

**Table 5.5** UC/PC-wise net sown land and its ratio to the total respective area

| | 1989 | | 1991 | | 1996 | | 2000 | | 2005 | |
|---|---|---|---|---|---|---|---|---|---|---|
| UC/PC | Cultivated | % of total | Cultivated | % of total | Cultivated | % of total | Cultivated | % of total | Cultivated | % of total |
| D. Khanzai | 1,489.3 | 18.4 | 3,602.6 | 44.4 | 4,712.9 | 58 | 2,542.2 | 31.1 | 3,317.4 | 40.9 |
| Batezai | 1,782.3 | 10.8 | 2,725.7 | 16.5 | 3,383.6 | 20.4 | 2,074.4 | 12.5 | 3,636.5 | 22 |
| Gangalzai | 262.4 | 5 | 652.7 | 12.5 | 809.4 | 15.5 | 665.5 | 12.8 | 1,006.27 | 19.3 |
| Huramzai | 1,195.8 | 10.8 | 1,786.8 | 16.2 | 2,251.6 | 20.4 | 1,877.2 | 17 | 2,531.1 | 22.9 |
| Malakyar | 1,810 | 14.3 | 3,457.8 | 27.3 | 4,277.8 | 33.7 | 3,773.9 | 29.8 | 4,495 | 53.4 |
| Manzaki | 243.9 | 2 | 724.8 | 6 | 773.7 | 6.4 | 605.9 | 5 | 954.99 | 7.9 |
| Manzari | 955 | 6.7 | 2,486.9 | 17.4 | 3,063.1 | 21.5 | 1,643.9 | 11.5 | 2,865.1 | 20 |
| N. Malezai | 1,258.6 | 4.6 | 2,805.9 | 10.3 | 4,042.3 | 14.8 | 1,270.8 | 4.7 | 4,702.3 | 17.3 |
| Saranan | 36.7 | 0.2 | 80.2 | 0.3 | 192.6 | 0.8 | 18.7 | 0.1 | 255.1 | 1 |
| Simzai | 769.3 | 5.8 | 1,502.9 | 11.3 | 1,771 | 13.3 | 1,614.8 | 12 | 1,996.96 | 15 |
| Tora Shah | 2,273 | 16.2 | 3,721.5 | 26.5 | 4,292.6 | 30.6 | 2,809.3 | 20 | 3,736.8 | 26.6 |
| Yaro | 89.5 | 0.7 | 157.3 | 1.3 | 172.5 | 1.4 | 69.4 | 0.6 | 350.3 | 2.9 |
| Alizai | 330.4 | 3.8 | 741.3 | 8.6 | 803.6 | 9.4 | 516.9 | 6 | 789.66 | 9.2 |
| Gulistan | 8,305.3 | 14.4 | 11,979 | 20.8 | 14,800.9 | 25.7 | 11,401.3 | 19.8 | 15,924.9 | 27.6 |
| Maizai | 2,188.9 | 13.8 | 2,869.7 | 18.1 | 3,087.7 | 19.5 | 3,040.5 | 19.2 | 3,254.38 | 20.6 |
| Q. Abdullah | 3,457.6 | 10.4 | 5,915.4 | 17.8 | 6,218.7 | 18.7 | 3,962.5 | 12 | 4,358 | 13.1 |
| Segi | 857.2 | 0.9 | 2,732 | 3 | 3,444.5 | 3.8 | 2,229 | 2.5 | 6,529.6 | 7.2 |

Source: Statistics derived from the Landsat scenes

**Table 5.6** Percentage frequency of farm-fallowing farmers by fallowed land ratio, 2008

| | Kharif season (summer) | | | | | Rabi season (winter) | | | | |
|---|---|---|---|---|---|---|---|---|---|---|
| | % of fallowing farmers | % ratio of fallowed land in under-cultivation land | | | | % of fallowing farmers | % ratio of fallowed land in under-cultivation land | | | |
| UC/PC | | 1–20 % | 21–40 % | 41–60 % | 61–80 % | | 1–20 % | 21–40 % | 41–60 % | 61–80 % |
| Yaro | 66.7 | 33.3 | 16.7 | 16.7 | 33.3 | 33.3 | – | 33.3 | 33.3 | 33.3 |
| Batezai | 37.5 | 100 | – | – | – | 12.5 | 100 | – | – | – |
| Manzari | 28.6 | 20 | – | – | 50 | 28.6 | 50 | – | – | 50 |
| Malakyar | – | – | – | – | – | – | – | – | – | – |
| Torashah | 18.8 | 33.3 | 33.3 | – | 33.3 | – | – | – | – | – |
| Saranan | 42.9 | – | – | 33.3 | 66.7 | 57.1 | – | 50 | 25 | 25 |
| Manzaki | – | – | – | – | – | – | – | – | – | – |
| D. Khanzai | 66.7 | – | 44.4 | 44.4 | 11.1 | 41.7 | 20 | 20 | 60 | – |
| N. Malezai | 20 | 50 | 50 | – | – | 20 | 100 | – | – | – |
| Huramzai | 50 | 66.7 | – | 33.3 | 3 | 41.7 | 20 | 60 | 20 | – |
| Gangalzai | – | – | – | – | – | – | – | – | – | – |
| Simzai | – | – | – | – | – | – | – | – | – | – |
| Alizai | 55.6 | 60 | 60 | – | – | 33.3 | 100 | – | – | – |
| Gulistan | 40 | 16.7 | 50 | 33.3 | – | 33.3 | 40 | 20 | 40 | – |
| Q. Abdullah | 42.9 | 33.3 | 66.7 | – | – | 28.6 | 100 | – | – | – |
| Maizai | 18.2 | 100 | – | – | – | 100 | 100 | – | – | – |
| Segi | 55.8 | 41.7 | 29.2 | 29.2 | – | 39.8 | 58.4 | 33.3 | 25 | – |
| **Gross mean %age** | **32** | **32.6** | **20.6** | **11.2** | **11.6** | **27.6** | **40.5** | **12.7** | **12** | **6.4** |

Source: Field Survey

though horticulture farms are considered to be at 100 % intensity every season, most of them are considered as kharif crops as they fruit in that season; in the rabi season, they just survive as dormant trees.

In terms of the amount of fallow land, the two cropping seasons are discussed separately hereunder:

### *Kharif Season*

Of the total 32 % of farm-fallowing farmers in this season, most (32.6 %) fallowed only 20 % or less of under-cultivation land, i.e., their farms were at ≥80 % farm intensity; 20.6 % farmers fallowed between 21 % and 40 %, meaning that farm intensity there was 60–79 % (Table 5.6).

### *Rabi Season*

A total 27.6 % of farmers reported farm fallowing in this season. Of these, 40.5 % fallowed ≤20 % of their lands. This means a farm intensity of ≥80 %. The other 12.7 % of the farmers fallowed 21–40 % of their lands, giving 60–79 % farm intensity (Table 5.6).

Table 5.7 UC/PC-wise compelling factors of farm fallowing

| | Frequency of reported factors (%) | | | |
|---|---|---|---|---|
| UC/PC | Water scarcity | Low soil fertility | Load shedding | Frost |
| Yaro | 45.5 | 27.3 | 9.1 | 18.2 |
| Batezai | 60 | 20 | – | 20 |
| Manzarai | 50 | – | – | 50 |
| Malakyar | – | – | – | – |
| Torashah | 50 | 50 | – | – |
| Saranan | 33.3 | 33.3 | 11.1 | 22.2 |
| Manzaki | – | – | – | – |
| D. Khanzi | 75 | 16.7 | – | 8.3 |
| N. Malezai | 66.7 | – | – | 33.3 |
| Huramzai | 50 | 25 | 8.3 | 8.4 |
| Gangalzi | – | – | – | – |
| Samzai | – | – | – | – |
| Alizai | 80 | 20 | – | – |
| Gulistan | 45.5 | 18.2 | 18.2 | 9.1 |
| Q. Abdullah | 50 | 50 | – | – |
| Maizai | 66.7 | – | 33.3 | – |
| Segi | 57.9 | 15.8 | 10.5 | 15.8 |

Source: Field Survey

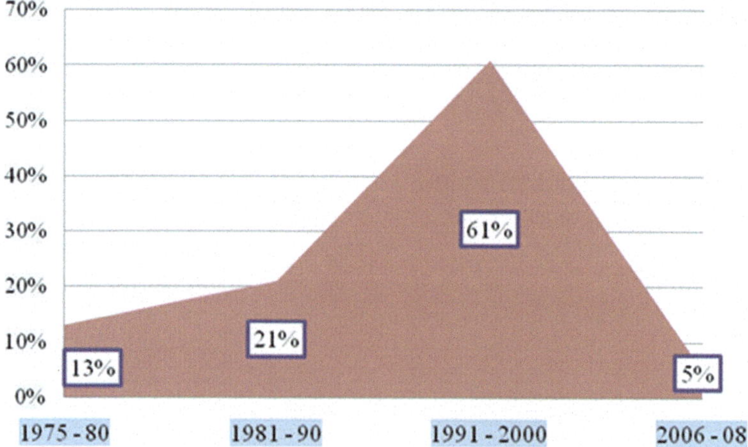


Fig. 5.4 Percentage of respondents by time when they started farm fallowing (Source: Field Survey)

The survey asked for the year in which farmers were compelled to start the practice of farm fallowing due to water scarcity or other reasons given in Table 5.7. The years reported were grouped, albeit with irregular intervals, for graphic presentation (Fig. 5.4). The figure indicates that the practice of farm fallowing started

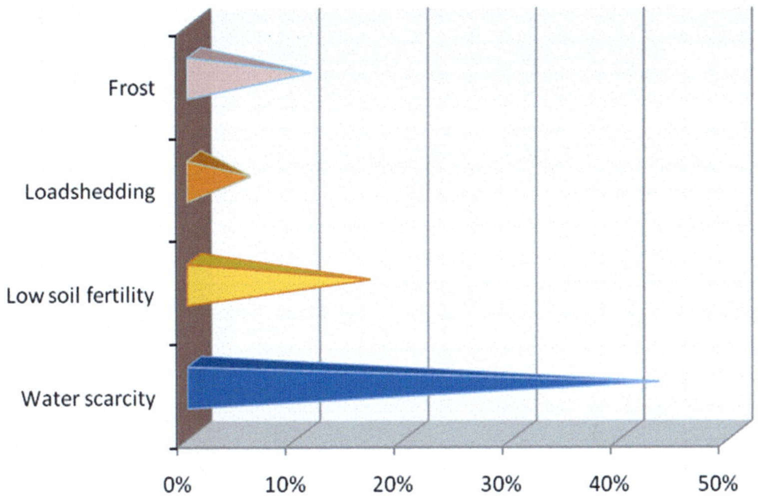


**Fig. 5.5** Modal frequencies of the compelling factors of farm fallowing (Source: Field Survey)

since 1975 and the number of farmers involved in that practice gradually increased through to the year 2000. From there until the end of 2005, no new cases were reported. A small number (5 %) of farmers practicing farm fallowing appeared again during 2006–08.

The overwhelming reason for farm fallowing was water scarcity (43 %). Others included lack of soil fertility (16.3 %), frost in rabi season (10.9 %), and electricity load-shedding (5.3 %) (Fig. 5.5).

### Intercropping

Intercropping means the growing of minor crops in the empty spaces between major crops, simultaneously in the same field. This practice is adopted in many farming communities of the world, mostly in fruit orchards; especially when the orchard is new and the trees have not yet reached full maturity. Since trees are planted in rows with enough inter-plant distance to suit their peak canopy and root system, the spaces can be used to grow a variety of small size crops therein. Though intercropping takes farm intensity to above 100 %, it does also require heavy fertilizer and irrigation input.

Although the study area is mostly horticulture, intercropping was very rarely found. Only a few respondents reported practicing it in kharif season in four UCs of Pishin district; there were no cases in rabi season (Table 5.8). The main deterring reasons were irrigation and soil fertility limitations. It is also meaningful to note that all the four UCs where intercropping was reported are located in/or close to the northern piedmont zone of the valley, which has several advantages in aquifer characteristics (e.g., it is close to a number of DADs, and is the first to receive surface streams).

**Table 5.8** Percentage frequency of farmers by intercropped land and the crops involved

| | Ratios of land and farmers in intercropping practice | | | Crops intercropped | |
|---|---|---|---|---|---|
| | Kharif season | | | Kharif season | |
| Union council | 10 % of cultivated land intercropped | 20 % of cultivated land intercropped | UC total | Onion in orchard farm | Potato in orchard farm |
| Malakyar | 9.1 | 9.1 | 18.2 | 100 | – |
| Torashah | 6.3 | 6.3 | 12.6 | 50 | 50 |
| Simzai | 25 | – | 25 | – | 100 |
| Gangalzai | – | 18.2 | 18.2 | 50 | 50 |

Source: Field Survey

Table 5.8 shows that, while on one hand, the farmers practicing intercropping are few, on the other hand, the ratio of land they put to intercropping is also little – between 10 and 20 %. The other point to note is that onion and potato crops are traditionally cultivated inside apple orchards, which is in fact the most ubiquitous horticulture here.

### 5.2.2 *Impacts on Cropping Pattern*

The study area has historically been the hub of fruit orchards in Balochistan. The majority are apple orchards, followed by grape, apricot, plum, and pomegranate. Apple is a high-delta crop, i.e., it requires a comparatively greater level of irrigation. During the drought of 1998–2004, and the general drawdown of aquifer, a great number of farmers lost their fruit orchards.

We know that fruit cultivation is a long-term business. Once lost, an orchard requires several years to rehabilitate to fruiting. Due to aridity and aquifer deterioration, the farmers are realizing risks in horticulture and are, therefore, gradually transforming from horticulture to field crops such as vegetables, tobacco, wheat, etc. Field crops are annual plants; therefore, damage inflicted by one or a few harsh year(s) would affect the farmers' income only during those bad years. However, aquifer deterioration, if it continues, will ultimately grossly reduce production, and shall affect both subsistence crops and those produced commercially.

As a whole, 62 % of the farmers have made alterations in kharif season crops and 51.5 % in rabi season crops, mainly due to irrigation scarcity (Table 5.9). In lieu of horticulture, in kharif season, 58.8 % of the farmers opting to change are cultivating vegetables, 32.8 % tobacco only, 5.2 % tobacco and vegetables mixed, 2 % grape instead of apple (grape is a low-delta crop), and others are cultivating a mixture of grapes, vegetables and melons. In rabi season, wheat has been adopted widely in lieu of horticulture – especially the apple orchards. Note that a fruit farm is counted in both kharif and rabi seasons, as fruit trees are perennial plants. Table 5.9 shows that spatial variations regarding cropping pattern changes exist from locality to locality within Pishin Valley.

**Table 5.9** Water scarcity-driven changes in crop selection, 2008

| | Kharif season (%) | | | | | | | Rabi season (%) | | | | |
|---|---|---|---|---|---|---|---|---|---|---|---|---|
| | | Change ratio of cultivated land in the change adopting farmers | | | | | | | Change ratio of cultivated land in the change adopting farmers | | | |
| UC/PC | No change farmers | Orchard to veg. | Orchard to tobacco | Orchard to tobacco and veg. | Apple to grapes | Apple to grapes and veg. | Orchard to veg. and melons | No change farmers | Orchard to wheat | Orchard to veg. | Apple to grapes | Apple to grapes and wheat |
| Yaro | 88.9 | – | 100 | – | – | – | – | 100 | – | – | – | – |
| Batezai | 25 | 50 | 50 | – | – | – | – | 62.5 | 33.3 | 66.7 | – | – |
| Manzarai | 42.9 | 100 | – | – | – | – | – | 42.9 | 100 | – | – | – |
| Malakyar | 63.6 | 75 | – | 25 | | – | – | 72.7 | 66.7 | 33.3 | – | – |
| Torashah | 25 | 41.7 | 16.7 | 16.7 | 25 | – | – | 56.3 | 100 | – | – | – |
| Saranan | 14.3 | – | 100 | – | – | – | – | 42.9 | 100 | – | – | – |
| Manzaki | 66.7 | – | 100 | – | – | – | – | 33.3 | 50 | 50 | – | – |
| D. Khanzai | 16.7 | 90 | 10 | – | – | – | – | 16.7 | 100 | – | – | – |
| N. Malezai | 20 | – | 87.5 | 12.5 | – | – | – | 100 | 0 | – | – | – |
| Huramzai | 8.3 | 45.5 | 9.1 | 27.3 | 9.1 | 9.1 | – | 16.7 | 71.4 | – | 11.1 | 11.1 |
| Gangalzi | 36.4 | 100 | – | – | – | – | – | 45.5 | 100 | – | – | – |
| Samzai | 37.5 | 80 | 20 | – | – | – | – | 37.5 | 60 | 40 | – | – |
| Alizai | 33.3 | 66.7 | 33.3 | – | – | – | – | 55.6 | 100 | – | – | – |
| Gulistan | 13.3 | 92.3 | – | 7.7 | – | – | 7.7 | 20 | 100 | – | – | – |
| Q. Abdullah | 28.6 | 80 | 20 | – | – | – | – | 42.9 | 100 | – | – | – |
| Maizai | 72.7 | 100 | | – | – | – | – | 72.7 | 100 | – | – | – |
| Segi | 52.6 | 77.8 | 11.1 | – | – | – | 11.1 | 57.9 | 100 | – | | – |
| G. Avg. | 38 | 58.8 | 32.8 | 5.2 | 2 | 0.5 | 1.1 | 51.5 | 75.4 | 11.2 | 0.7 | 0.7 |

Source: Field Survey

### 5.2.3 *Economic Efficiency of Tubewell Farming*

Economic efficiency means the ratio between the cost of inputs and the value of the return. Poverty has been among the major concerns in human societies. As Pakistan is an agricultural economy, poverty is directly linked to agricultural development in every corner of the country. To this end, positive efficiency of irrigated farming should be the vehicle to take the country's masses above the poverty line. This is also one of the Millennium Development Goals through ensuring sustainable development.

Agriculture requires a number of inputs, including machinery, seeds, fertilizer, and irrigation, etc. However, our survey was limited to finding data on the cost of irrigation accessories only, as it is the prime factor/input and not only has a key role in net farm profitability but also could even determine the very existence of agriculture in the study area.

#### Irrigation Cost

The following three factors account for the majority of irrigation costs:

- Initial capital cost of tubewell installation, including drilling and casing.
- Cost of the prime mover (pumping motor), along with essential piping and wiring.
- Recurring expenditure for O & M, including electricity tariff and repair bills.

The cost of irrigation is calculated in terms of the local currency – Pakistani rupee (exchange rate in April 2014 was approximately $US1 for 98 Pakistani rupees). Although, devaluation of this currency occurs from time to time, in real international terms, the price of irrigation accessories has been rising rapidly.

##### Cost of Tubewell Installation

The initial capital cost for tubewell installation (excluding the cost of the prime mover) varies among localities in the valley, depending upon factors such as depth to the watertable, boring depth, brand of material, etc. During 1981–90, the majority (40.3 %) of the tubewells had an initial capital cost up to only 50,000 rupees (Table 5.10); whereas the maximum limit of that cost was 3 lac and 50 thousand rupees (Rs. 350,000); however, only 0.7 % of the tubewells were at this upper limit of the cost. During 1991–2000, the initial capital cost per tubewell became 50,000 to 1 lac rupees; however, only 21.8 % of the tubewells were at this extreme limit, and it was the second highest number in this period. Since the year 2000, the initial capital cost of tubewells escalated in leaps and bounds. For instance, 44 % of the tubewells were installed at the cost of 5–6 lac rupees during 2001–04; and at above 8 lac rupees in the next 4-year span of 2005–08 (Table 5.10).

The rapid increase in the initial capital cost of tubewell installation is primarily due to watertable depletion. Apart from the immense cost, another worrisome issue

**Table 5.10** Percentage frequency of tubewells by initial capital cost (Pak. rupees)

| Year | Cost | | | | | | |
|---|---|---|---|---|---|---|---|
| | ≤50 Th[a] | 51 Th–1 Lac[b] | 101 Th–150 Th | 151 Th–2 Lac | 201 Th–250 Th | 251 Th–3 Lac | 301 Th–350 Th |
| 1981–90 | 40.3 | 17.5 | 18.8 | 20.8 | 0.7 | 1.3 | 0.7 |
| 1991–2000 | 1.2 | 26.7 | 5.5 | 12.1 | 7.3 | 17 | 21.8 |
| 2001–04 | – | – | – | 7.9 | 0.6 | 1.8 | 1.2 |
| 2005–08 | – | – | – | – | – | – | 0.7 |

| Year | Cost | | | | | | |
|---|---|---|---|---|---|---|---|
| | 351 Th–4 Lac | 401 Th–450 Th | 451 Th–5 Lac | 501 Th–6 Lac | 601 Th–7 Lac | 701 Th–8 Lac | > 8 Lac |
| 1981–90 | – | – | – | – | – | – | – |
| 1991–2000 | 6.1 | – | 0.6 | 1.2 | – | 0.6 | – |
| 2001–04 | 4.9 | 7.3 | 14 | 44 | 17 | 1.2 | – |
| 2005–08 | 7.9 | 0.7 | 7.3 | 15.9 | 25.8 | 35.8 | 6 |

Source: Field Survey
[a]Th = thousand
[b]1 Lac = 1 hundred thousand

**Table 5.11** Percentage frequency of tubewells by size of the prime mover (horsepower)

| Time | Horsepower 20 | 25 | 30 | 35 | 40 | 45 | 50 |
|---|---|---|---|---|---|---|---|
| 1981–90 | 39 | 46 | 12.8 | – | 2 | – | – |
| 1991–2000 | 3 | 14.5 | 61.8 | 19.4 | 1.2 | – | – |
| 2001–04 | – | – | 34.4 | 16.6 | 49 | – | – |
| 2005–08 | – | – | 7.2 | 11 | 73.2 | – | 8.5 |

Source: Field Survey

is the drying out of tubewells after short periods of service because, in most cases, the watertable declined at a rate faster than farmers expected or were prepared for.

### Cost of Pumping Machine

Because of ubiquitous electrification, the Government electricity tariff subsidy (FRP), and opportunities to steal electricity, etc., all the pumping machines used by the sample farmers were electric. Although their prices also increased with general inflation, here there is also another facet to the issue. With depletion of the watertable, more and more powerful motors were required to abstract water. Starting from 1981 to 90, model types were some 20–25 hp (Table 5.11), but as the watertable dropped further, 30 hp motors came into use during 1991–2000; and 40 hp motors since then. Though few in number, 50 hp motors were also in use towards the end of our data period. Obviously, a high-power machine is more expensive than a low power machine in terms of several variables (e.g., shelf cost, repair cost, and consumption of electricity).

**Table 5.12** Percentage frequency of tubewells by cost of the prime mover (×1,000 rupees)

| | Cost | | | | | | | |
|---|---|---|---|---|---|---|---|---|
| Time | 10–20 | 21–30 | 31–40 | 41–50 | 51–60 | 61–70 | 71–80 | >81 |
| 1981–90 | 41.4 | 42 | 16.4 | – | – | – | – | – |
| 1991–2000 | 6.8 | 26.7 | 24.8 | 41.6 | – | – | – | – |
| 2001–04 | – | 2.5 | 15 | 47.8 | 34.6 | – | – | – |
| 2005–08 | – | – | 0.7 | 8.7 | 65.8 | 21.5 | 2 | 1.3 |

Source: Field Survey

**Table 5.13** Percentage ratio of tubewells by per month O & M expenditure

| | Expenditure | | | | | | | | |
|---|---|---|---|---|---|---|---|---|---|
| Time | ≤1 Th[a] | 11 H[b]–2 Th | 21 H–3 Th | 31 H–4 Th | 41 H–5 Th | 51 H–6 Th | 61 H–7 Th | 71 H–8 Th | >8 Th |
| 1981–90 | 5.2 | 69 | 23.9 | 0.6 | 0.6 | – | – | – | – |
| 1991–2000 | – | 18.2 | 26.7 | 50.9 | 1.8 | 1.2 | 0 | 1.2 | – |
| 2001–04 | – | 1.8 | 7.4 | 33.7 | 42.9 | 43.6 | 1.2 | 2.5 | – |
| 2005–08 | – | – | 5.2 | 13.7 | 20.3 | 41.8 | 11.8 | 3.9 | 3.3 |

Source: Field Survey
[a]Th = thousand Pak. rupees
[b]H = hundred Pak. rupees

Apart from the horsepower size, the shelf cost of motors also varies from brand to brand. It was noted that during 1981–90, about 42 % of the motors were purchased at around 25, 000 rupees (Table 5.12); and ultimately in 2005–08, the prices went up to about 1 lac rupees per head.

### Cost of Tubewell O & M

By cost of tubewell operation we mean the electricity tariff only; and the cost of tubewell maintenance refers here to all types of repair works, including motor rewinding, pipe system repair, etc. Table 5.13 shows that, like other expenditures, the O & M cost as a whole has risen throughout our study period. Nonetheless, for many tubewells, the operation cost – electricity tariff – is fixed (Rs. 4,000/month) as per the flat rate policy, adopted since the 1970s. However, in some cases, the subsidy is not allowed and those tubewells are called 'unsubsidized tubewells'.

Table 5.13 shows that during 1981–90, the most frequent cost range was between 1 and 2 thousand rupees per month, whereas during 1991–2000, it was 3–4 thousand rupees a month. Ultimately during 2005–08, the O & M cost had increased to 6,000 rupees in most cases, and even to 8,000 rupees in some others.

**Table 5.14** Ratio of farmers by irrigation ownership and type of pumping machine

| | Water ownership (%) | | Type of pumping machine (%) | |
|---|---|---|---|---|
| Time | Owner | Buyer | Electric | Diesel |
| 1981–90 | 97.2 | 2.8 | 100 | 0 |
| 1991–2000 | do | do | do | 0 |
| 2001–04 | do | do | do | 0 |
| 2005–08 | 96.1 | 3.9 | do | 0 |

Source: Field Survey

**Profitability of Tubewell Farming**

The literature review suggested that the study area possesses farmers who do not have their own tubewells, but rather buy irrigation water, and that diesel-operated tubewells do exist in the valley. Therefore, in the survey questionnaire, we incorporated questions directed towards both water owner and buyer farmers; and both electricity-operated and diesel-operated tubewell cases. However, in the field we did not find any diesel-operated tubewells (Table 5.14). Regarding the water owner and buyer aspect, our sample did include a few water buyers, but their ratio in the overall sample was too small (2.8–3.9 %) to produce any significant statistics and results (Table 5.14). Therefore, our analysis regarding farm profitability is based entirely on the owner case scenario of electricity-operated tubewells.

Gross Farm Income

Gross farm income means the market value of the produce, irrespective of the cost of production. Farmers generally showed sensitivity and reluctance to the question of income; therefore, quantum variation is found in the reported figures, depending upon the perceptions and psyche of respondents. The farmers generally try to show themselves as devastated. Further, the nature of soil, types of crops, size of the household (as to whether it can spare produce for market or consumes the whole itself), etc. also play their role. Regardless, the gross farm income was from less than 5,000–20,000 rupees per acre per annum in most (70 %) cases during 1981–90; in the other 30 %, it varied between 20,000 and 60,000 rupees with a decreasing frequency of farmers in increasing income groups (Table 5.15). Ten years later, during 1991–2000, the gross farm income in monetary terms shows a slight increase. For 45.3 % it remained in the range of 5,000–20,000 rupees, but for the rest it was above 20,000 to 1 lac rupees. In this period, a great number of farmers (12 %) reported per acre per annum gross farm income as much as 1 lac rupees. It is interesting to note that the 1 lac rupees per acre per annum margin was achieved in 1991–2000, and by as many as 12 % of the farmers, but onwards till 2008, none attained this benchmark. Rather, the percentage ratio of farmers at this level of gross

**Table 5.15** Percent ratio of farmers by per acre/per annum gross farm income

| Time | Income | | | | | | | |
|---|---|---|---|---|---|---|---|---|
| | ≤5,000 | 5,001–10 Th | 1,000–15 Th | 15,001–20 Th | 20,001–25 Th | 25,001–30 Th | 30,001–35 Th | 35,001–40 Th |
| 1981–90 | 24.5 | 22.5 | 13.8 | 20 | 6.9 | 6.8 | 1.4 | 2.7 |
| 1991–2000 | 7.4 | 17.4 | 11.4 | 9 | 2.3 | 3.4 | 4.5 | 2 |
| 2001–04 | 4 | 8.9 | 12 | 16 | 1.7 | 6.4 | 3 | 1.4 |
| 2005–08 | 4.3 | 6.5 | 10.3 | 17.6 | 6 | 3.3 | 1.4 | 8 |
| Time | Income | | | | | | | |
| | 40,001–45 Th | 45,001–50 Th | 50,001–60 H | 60,001–70 Th | 70,001–80 Th | 80,001–90 Th | 90,001–1 Lac | >1 Lac |
| 1981–90 | 0.5 | 0.4 | 0.5 | – | – | – | – | – |
| 1991–2000 | 1.8 | 2.7 | 4.3 | 6.1 | 3.6 | 3.8 | 3.6 | 12.4 |
| 2001–04 | 2 | 11 | 4 | 4.9 | 7.7 | 9.6 | 4 | 3.4 |
| 2005–08 | 2.2 | 7.2 | 5 | 6.8 | 6 | 0.8 | 3.5 | 2.4 |

Source: Field Survey

farm income has sharply dropped to 3.4 % during 2001–04, and later to 2.4 % during 2005–08. Table 5.15 also shows that from 1981 until 2008, the gross farm income of a number of farmers did not progress from ≤5,000 rupees. Keeping in view the continuously increasing inflation rate, this means that farm income in the study area, in gross terms too, has generally not increased.

## Net Farm Income

Net income means the sale price of a product minus the cost of its production; what we call 'profit' in simple terms. The net farm income situation is not much different from that of gross farm income. Like gross income, the net income report also varies greatly from farmer to farmer and area to area, depending upon several factors, including social and political reservations of the farmers while showing own wealth, farming and marketing skills of the farmers, farm size, quality of the soil, household size, and common-income or otherwise living culture, etc.

From the scenario of 1981–90 (Table 5.16), it appears that nearly half of the farmers (44.1 %) had a net farm income of up to 5,000 rupees or less per acre per annum; and the maximum derived by any farmer was 30,000 rupees or less per acre per annum. During 1991–2000, the overall net farm income range is wider by far. There are a great number of farmers (17 %) at the income level of 5,000 rupees or less, as well as a few (1.4 %) with a net farm income as high as 1 lac rupees. Further, during this period, besides the first two income groups (≤5,000–10,000 rupees), the frequencies of farmers in the net farm income groups vary rather smoothly. For economic sustainability of the area's agricultural economy, it is alarming to note that the net farm income range shrunk from the maximum level of 1 lac rupees in the years since 2000.

Further, it can be seen that 44.1 % of farmers were at the minimum net farm income level of ≤5,000 rupees. Later, during 1991–2000, this ratio sharply dropped

**Table 5.16** Percent ratio of farmers by per acre/per annum net farm income

| Time | Income | | | | | | | |
|---|---|---|---|---|---|---|---|---|
| | ≤5,000 | 5,001–10,000 | 10,001–15,000 | 15,001–20,000 | 20,001–25,000 | 25,001–30,000 | 30,001–35,000 | 35,001–40,000 |
| 1981–90 | 44 | 13.2 | 19.8 | 9 | 6.2 | 2.7 | – | – |
| 1991–2000 | 17.5 | 24 | 6.3 | 4 | 5 | 2.8 | 6 | 6.8 |
| 2001–04 | 19.2 | 28.3 | 2.2 | 12.4 | 5.4 | 7 | 5.2 | 8.7 |
| 2005–08 | 23.4 | 31.5 | 12 | 12 | 7 | 7.2 | 4.3 | 0.5 |
| Time | Income | | | | | | | |
| | 40,001–45,000 | 45,001–50,000 | 50,001–60 Th | 60,001–70 Th | 70,001–80 Th | 80,001–90 Th | 90,001–1 Lac | >1 Lac |
| 1981–90 | – | – | – | – | – | – | – | – |
| 1991–2000 | 1.7 | 4 | 7.4 | 4 | 6.8 | 2.5 | – | 1.4 |
| 2001–04 | 4.4 | 5.4 | 1.3 | 1.6 | 1.3 | – | – | – |
| 2005–08 | – | 1.6 | 1.2 | 1 | – | – | – | – |

Source: Field Survey

to 17.5 %, but in the years ahead, it is rising continuously to 19.2 %, followed by 23.4 %. This means, until the year 2000, net farm income was on the rise, but it had been gradually slumping in the following periods. This may be why poverty is rooting deep in this rural society and people are showing more interest in jobs other than farming.

## 5.3 Summary

The agricultural enterprise of this region must struggle against many kinds of constraints. To assess its long-term sustainability, we used both social data (questionnaire) and instrumental data (Landsat). Analysis and discussion of those data was the subject of this chapter. Prior to the core concerns, this chapter provided a perspective on socio-economic information about the farmers. The descriptive statistics indicated large household sizes and widespread illiteracy, or low level of education, among the farmers. Driven by dwindling farm income, the majority of households are also conducting non-farm occupations as supplementary income to their annual budgets. Agriculture, though still showing growth trend, is greatly constrained by limitations of the aquifer as it is vital to farming in the dry climatic scenario of the area. The desertification phenomenon has started, and economic profitability has slumped in the face of immense depletion of the watertable. The current need is to increase farm intensity and practice intercropping to mitigate food shortages, but the analysis found a discouraging scenario in these regards, principally because of irrigation resource limitations. Changes in the cropping pattern have been noticed, both from high-delta to low-delta crops and from low-value to high-value crops.

# Chapter 6
# Results and Discussion: Part B

**Abstract** This chapter shares the same perspectives that drove the analysis in Chap. 5. However, while Chap. 5 took tubewell farming as a dependent variable being hampered by aquifer (independent variable) limitations, this chapter has undertaken the problem from the aspect of the aquifer potential being a dependent variable deteriorated by the progress of tubewell farming (independent variable). The conclusions drawn by both Chaps. 5 and 6 are in support of our hypothesis that though tubewell farming and the local aquifer are partners in serving society, unfortunately, they do not ensure sustainable existence of each other under the present traditions of action. To reduce the limitations of primary data obtained from the farmers on a long time-series basis through their unrecorded memories, three sets of cross-checking data were used in this study. While in Chap. 5, the land-use data collected through farmers was validated by satellite data, in this chapter the groundwater data is validated by educated views of agriculture- and irrigation-related expert officials, as well as by instrumental physico-chemical analyses of the groundwater.

**Keywords** Pishin Valley • Watertable • Aquifer • Field survey • Government role

## 6.1 Tubewell Farming Affecting Aquifer

Aquifers are typically the underground saturated regions of permeable rocks or unconsolidated materials (e.g. gravel, sand) from which can be extracted an economically feasible quantity of water using water wells. The surface (upper level) of saturated material in an aquifer is known as the 'watertable'. By aquifer potential we mean:

- The depth of the watertable from the surface of the earth.
- Economic viability and consequences of water extraction from the aquifer.
- Suitability of the quality of the groundwater for multiple uses, including human use and irrigation.

A.S. Khattak, *Mutual Sustainability of Tubewell Farming and Aquifers: Perspectives from Balochistan, Pakistan*, Advances in Asian Human-Environmental Research, DOI 10.1007/978-3-319-02804-0_6, 

**Table 6.1** Average DTW and per year fall of the watertable (ft)

| UC/PC | Depth 1981–90 | Depth 1991–2000 | Annual fall 1991–2000 | Depth 2001–04 | Annual fall 2001–04 | Depth 2005–08 | Annual fall 2005–08 |
|---|---|---|---|---|---|---|---|
| Yaro | 115 | 158 | 4.3 | 237 | 19.7 | 269 | 8 |
| Batezai | 80 | 144 | 6.4 | 212 | 17 | 256 | 11 |
| Manzarai | 75 | 75 | – | 296 | 55.2 | 332 | 9 |
| Malakyar | 120 | 215 | 9.5 | 295 | 20 | 395 | 25 |
| Torashah | 96 | 180 | 8.4 | 240 | 15 | 325 | 21.3 |
| Saranan | 66 | 77 | 1.1 | 96 | 4.8 | 111 | 3.8 |
| Manzaki | 117 | 217 | 10 | 410 | 48.2 | 420 | 2.5 |
| D. Khanzi | 75 | 136 | 6.1 | 208 | 18 | 354 | 36.5 |
| N. Malezai | 87 | 130 | 4.3 | 185 | 13.8 | 210 | 6.2 |
| Huramzai | 96 | 146 | 5 | 211 | 16.3 | 261 | 12.5 |
| Gangalzai | 77 | 166 | 8.9 | 289 | 30.8 | 348 | 14.8 |
| Simzai | 75 | 162 | 8.7 | 287 | 31.2 | 294 | 24.3 |
| Alizai | 92 | 180 | 8.8 | 219 | 9.8 | 353 | 33.5 |
| Gulistan | 82 | 82 | – | 302 | 55 | 350 | 12 |
| Q. Abdulah | 96 | 111 | 1.5 | 289 | 44.5 | 335 | 11.5 |
| Maizai | 80 | 93 | 1.3 | 284 | 47.8 | 348 | 16 |
| Segi | 80 | 130 | 5 | 225 | 23.8 | 283 | 14.5 |
| Average | 89 | 141 | 5.2 | 252 | 27 | 308 | 14 |

Source: Field Survey

### *6.1.1 Watertable Depletion*

The cost of groundwater pumping is directly proportional to its depth. Fall of the watertable is expressed by both statistical tables and graphics. The saturated zone was at an average depth of 89 ft during 1988–90 (Table 6.1). It declined to 141 ft during the next 10 years, at a rate of 5.2 ft per year. The fall of the watertable was phenomenal during 2001–04, with a maximum rate of 27 ft per year. This reflects effects of the prolonged and acute drought between 1998 and 2004. During 2005–08, the watertable fall rate was also great but it was less than in the previous period. Three factors may have caused this:

- The drought of 1998–04 ended in 2005 (Table 3.2). The relatively heavy precipitation in 2005 and onward may have contributed to the aquifer, or made the farmers less dependent on groundwater irrigation.
- From the year 2000 onward, the tubewells decreased in number due to drying out of many as a result of decreases in the watertable.
- With deepening of an aquifer, its ability to maintain a certain level of watertable increases, because it starts yielding less water per unit time of pumping.

Individual scenarios of the constituting UCs/PCs reveal disparity in the watertable depletion rates. Although, it has been established that the watertable has declined rapidly, in reality, it may not have declined so variedly from place to place

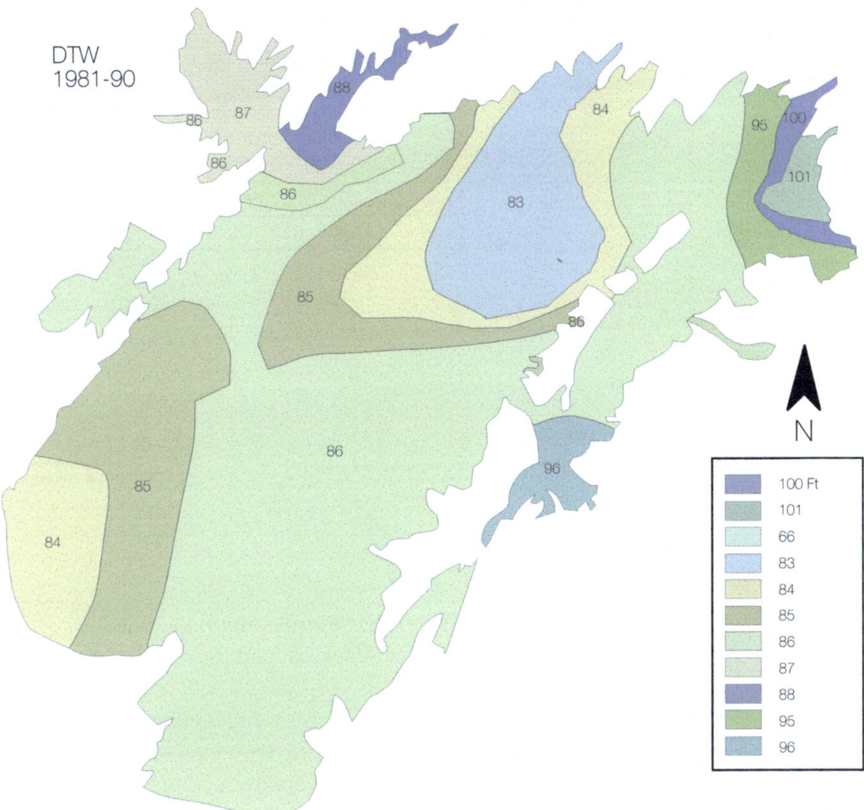


**Map 6.1** Average DTW (ft), 1981–90

within the valley. However, since we used reports from farmers, it is conceivable that their knowledge about the depth of the watertable may not have been precise, especially concerning levels in the distant past, and we had no alternative source of information available to us. However, the margin of error is reduced by averaging the reports. Generally, we note that relief of the land has a role in the depth of the watertable. For example, Maps 6.1, 6.2, 6.3, and 6.4 show that the watertable was relatively deeper in the surrounding piedmont flanks of the valley than in its central low-lying floor.

The maps showing the annual watertable fall rates (Maps 6.5, 6.6, and 6.7) show that the watertable depletion rate has always been faster in areas of dense tubewell-irrigated cultivation. However, there is one anomaly in this regard: the Ajram Shadezai area, located in the barren south of the valley, was not surveyed but for cartographic needs; to obtain complete coverage, it is included in the polygon of its neighboring surveyed area and given a uniform colour.

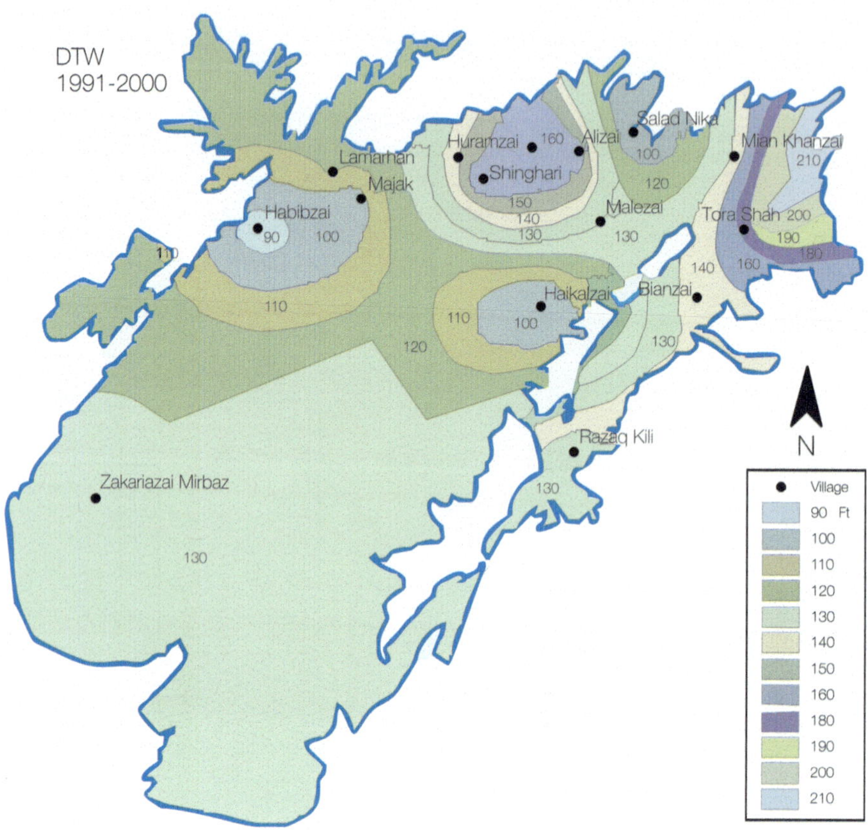


Map 6.2 Average DTW (ft), 1991–2000

On an individual basis among the sampled UCs/PCs, wherever there was high agricultural growth, there was also a high rate of watertable decline. For example, in the northward Manzari UC, the watertable dropped by 30 ft per year during 2005–08. Table 5.5 shows that net sown land in this area was almost twice as much in 2005 than it was in the year 2000 (satellite data). Similarly, the watertable declined by 20 ft annually at the N. Malezai UC, where cultivated area was nearly four times (from 4.66 to 17.25 % of the total area) bigger in 2005 than it was in the year 2000 (satellite data).

The field survey data, which are durational, show that in many parts of the valley, expansion of tubewell farming and the watertable are inversely related to one another; i.e. with greater expansion of tubewell-irrigated agriculture, more depletion of the aquifer has occurred.

Although, due to population spread, new lands have been brought under plough, new aquifer zones have been tapped, and agriculture on the whole has slowly but gradually expanded (Fig. 5.2), on the similar inverse pattern, the watertable has

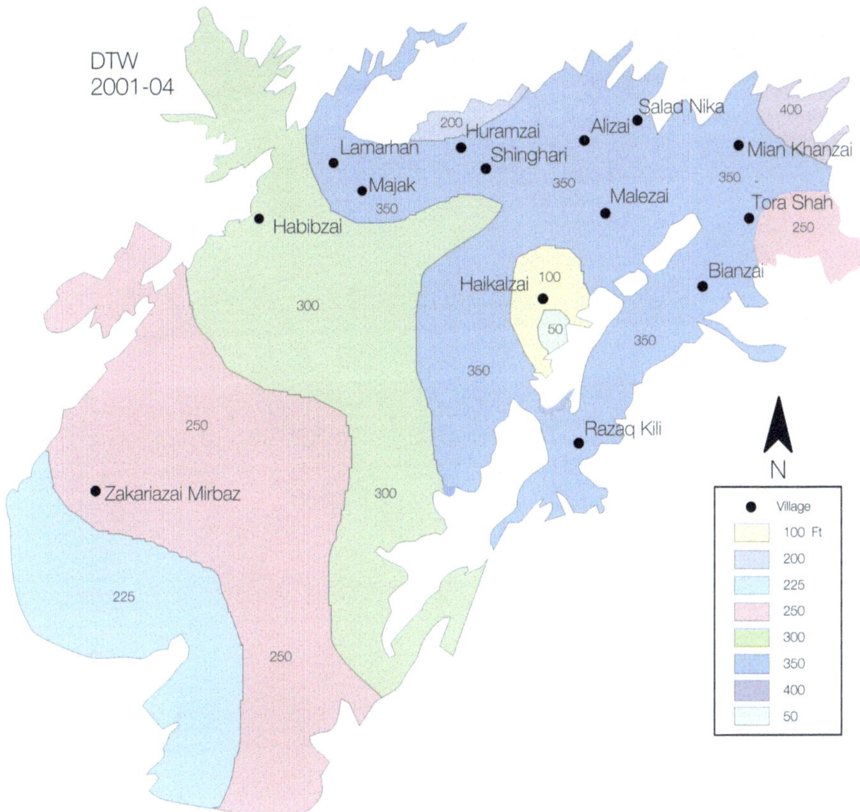


**Map 6.3** Average DTW (ft), 2001–04

always trended down since the starting time period for this study (1981). That is to say, the development of tubewell-irrigated agriculture has consistently resulted in the loss of aquifer potential because the aquifer's recharge potential has been weak, and this has not been addressed properly by policies and strategies. Thus, these facts support the basic idea of our hypothesis: that current agriculture and irrigation practices are unsustainable in the region.

### *6.1.2 Aquifer Mining*

The term 'aquifer mining' is used when an aquifer is exploited beyond a rechargeable depth; where it becomes non-renewable like ordinary minerals (e.g. petroleum, gas, etc.) within a rational time period. In the study area, the watertable has dropped

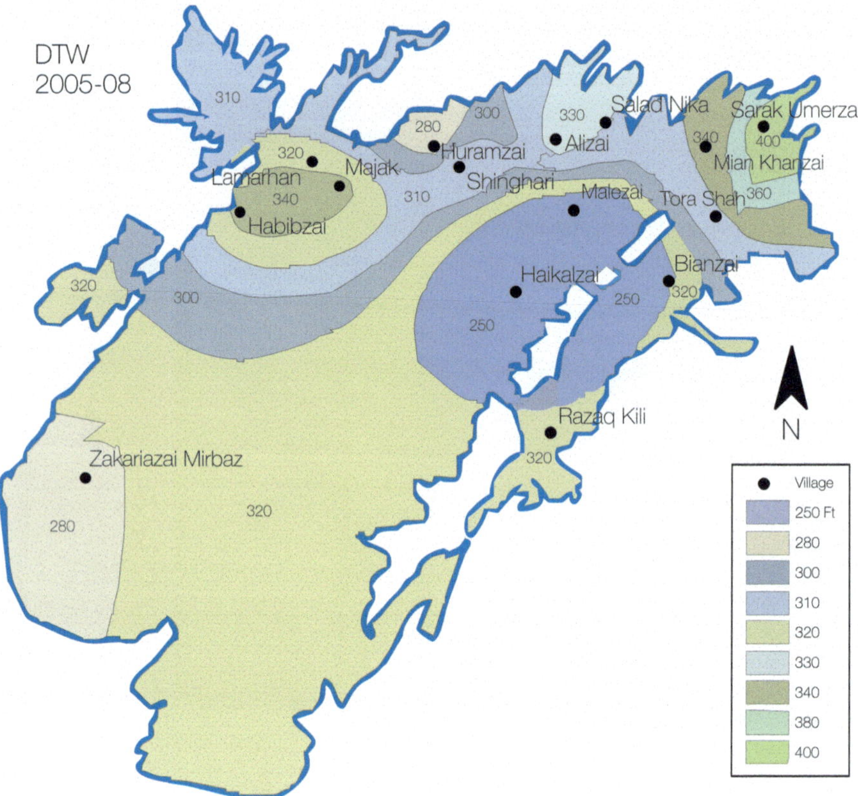


**Map 6.4** Average DTW (ft), 2005–08

too far (Table 6.1). Further, in the past, the farmers used to drill in the order of tens of feet into the saturation zone because it was considered more than enough for their irrigational needs over a long period (Table 6.2). However, with the rapid fall of the watertable, many older tubewells dried out (Table 6.4), resulting in immense losses for farmers. Now, farmers arrange finances from everywhere possible, but once they drill, they drill deeper until virtually the end of the saturation zone (Table 6.3). This means the aquifer is being mined, because the aquifer at hundreds of feet down in the earth cannot recharge by natural ways within a reasonable time span. Although, in some recent individual cases, tubewell bores have reached depths as far down as 800 ft, the average bore depth in the valley is 573 ft (Table 6.2); UC/PC-wise figures vary in this regard, depending primarily upon topographic irregularities.

In terms of period-to-period changes, Table 6.3 reveals that, in 1981–90, some 28.5 % of tubewell bores were no deeper than 100 ft. Obviously, when this was the drilling limit, the initial capital cost of tubewell installation was far less than it is now, when farmers are drilling to depths of near 1,000 ft. Apart from this, the experts

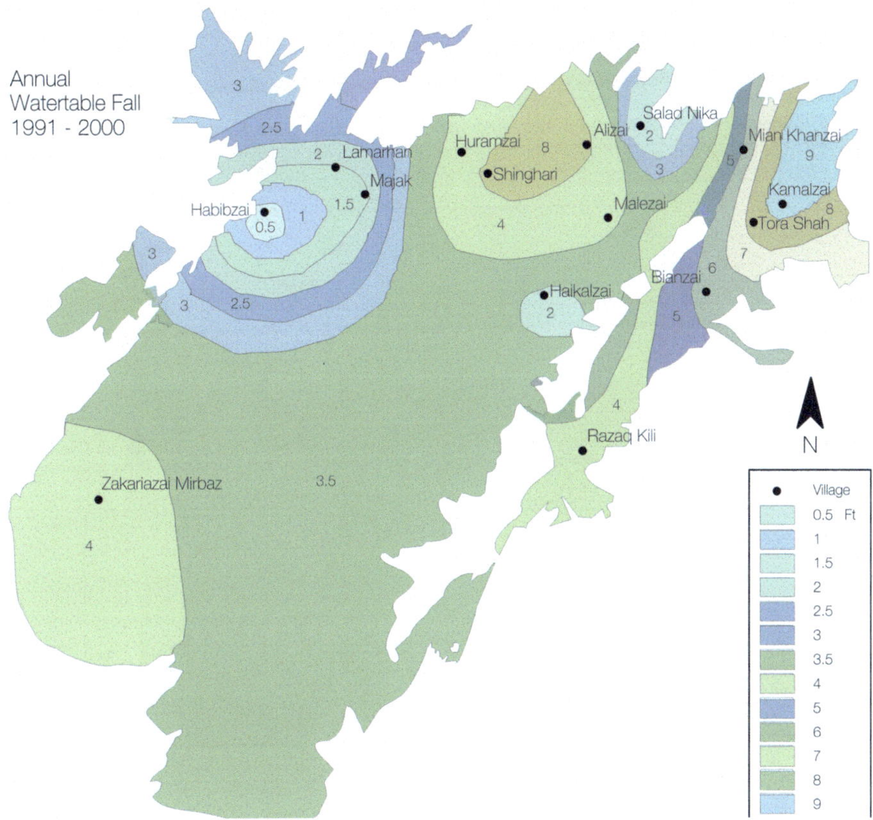


**Map 6.5** Per year watertable decline (ft), 1991–2000

suggest that as the drilling is going deeper it is coming closer to sea-level water, which is highly likely to generate the phenomenon known as 'seawater or saline water intrusion'. In fact, some careful instrumental analyses have indicated this phenomenon to have appeared in the groundwater resources of the study area.

Table 6.3 further shows that the culture of deeper drilling increased from the period 1981–2000. The explanation for that is the advent of the powerful drilling rigs, instead of the traditional bucket-type drilling machinery.

### *6.1.3 Reduction in Tubewell Discharge*

Such deep dropping of the watertable, and the consequent need for deeper drilling, has created social inequity. A great number of resource-poor farmers have been deprived of their tubewells (Tables 6.4 and 6.5), which is critical for

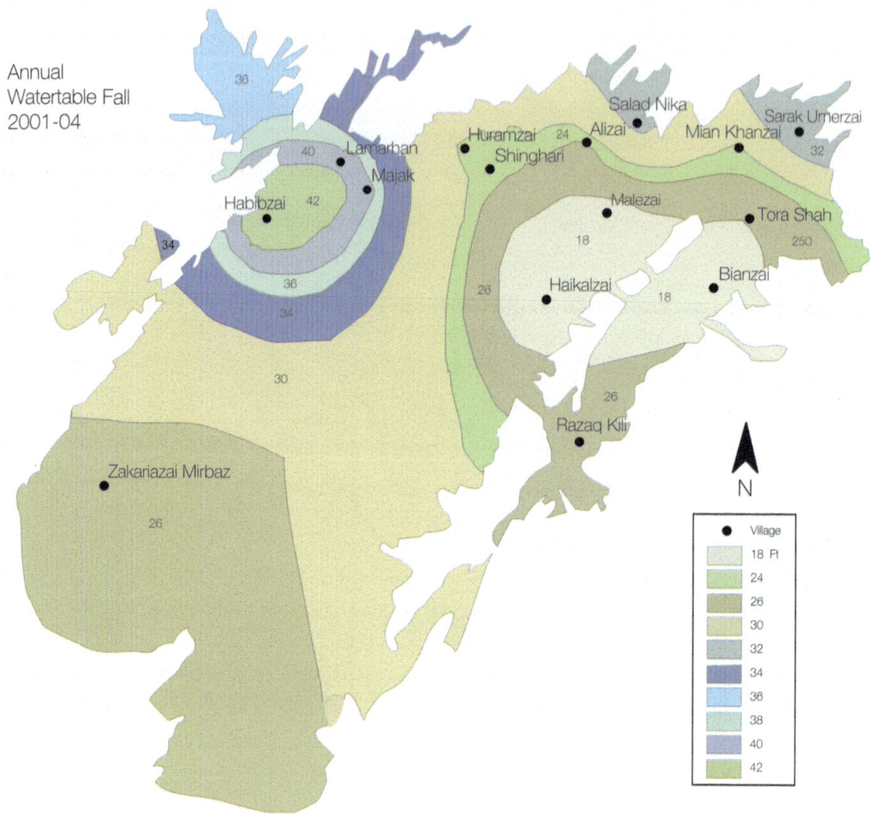


**Map 6.6** Per year watertable decline (ft), 2001–04

maintaining soil moisture in a dry region such as this. Table 6.5 shows that, during 1981–90, a total of 14.6 % of irrigation farmers did not have their own tubewells. They either depended upon karez irrigation (at that time, some karezes were alive in the area because the watertable was accessible by open digging), or shared water with others, per the area's traditional tenancy terms and conditions. In the next period (1991–2000), the proportion of tubewell-deprived farmers reduced sharply to 6.7 %, conceivably due to further electrification of the area and the Government's flat rate electricity policy for irrigation tubewells. Onward from the year 2000, the tubewell-deprived farmers consistently increased until 2008. Here, the cause may be the drying out of many older tubewells and the inability of the resource-poor farmers to install replacements in the face of the very high financial cost of deeper drilling.

Quite contrary to the resource-poor farmers, the well-off farmers were motivated to own a number of tubewells simultaneously because of the following two factors:

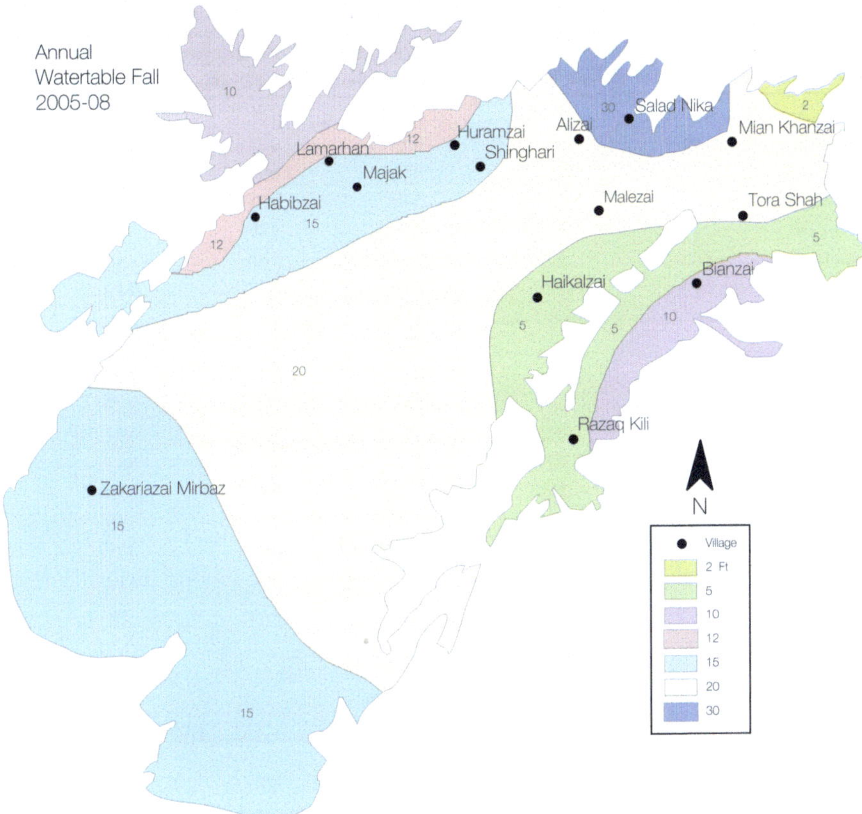


**Map 6.7** Per year watertable decline (ft), 2005–08

1. Discharge (water yield) per tubewell has reduced. This is because the deeper aquifers are more compressed by overlying material than are the shallow aquifers; hence, as the watertable is getting deeper, the discharge capacity of the tubewells decreases. Thus, the bigger and resourceful farmers solve this problem by installing many tubewells simultaneously. It is evident from Table 6.5 that, during 2001–05, some 50 % of farmers had two to five tubewells simultaneously. We even note, in a few cases during 2001–08, a proportion of farmers with 10–30 tubewells at one time. In owning many tubewells, the rich farmers also create another business; that is, the selling of water for irrigation to non-tubewell-owner farmers. They pay subsidized tariffs for the electricity consumed in their tubewells, but sell the water at a high profit margin.
2. The load-shedding in electricity supply is also a reason for wealthy farmers to install many tubewells. If they had one or a few tubewells, they will not be able to meet the water demands of their own farms, as well as those of their clients, because load-shedding allows them less time to operate tubewells. By having many tubewells simultaneously, they are able to compensate for the time lost through load-shedding.

**Table 6.2** Average depth of total boring for tubewells (ft)

| UC/PC | Bore depth 1981–90 | Bore depth 1991–2000 | Bore depth 2001–04 | Bore depth 2005–08 | Depth increase 1981–2008 |
|---|---|---|---|---|---|
| Yaro | 208 | 275 | 237 | 456 | 248 |
| Batezai | 168 | 237 | 559 | 425 | 257 |
| Manzarai | 75 | 111 | 686 | 696 | 621 |
| Malakyar | 325 | 370 | 540 | 575 | 250 |
| Torashah | 311 | 336 | 548 | 727 | 416 |
| Saranan | 125 | 367 | 404 | 568 | 443 |
| Manzaki | 217 | 425 | 630 | 637 | 420 |
| D. Khanzi | 247 | 286 | 458 | 643 | 396 |
| N. Malezai | 350 | 380 | 425 | 477 | 127 |
| Huramzai | 304 | 304 | 471 | 543 | 239 |
| Gangalzai | 280 | 280 | 443 | 509 | 229 |
| Simzai | 169 | 281 | 653 | 666 | 497 |
| Alizai | 167 | 369 | 611 | 750 | 583 |
| Gulistan | 102 | 235 | 592 | 652 | 550 |
| Q. Abdullah | 146 | 204 | 500 | 535 | 389 |
| Maizai | 89 | 139 | 480 | 516 | 427 |
| Segi | 212 | 235 | 504 | 624 | 412 |
| Average | 206 | 284 | 514 | 573 | 367 |

Source: Field Survey

The issue of load-shedding has several implications besides the possession of a number of tubewells by wealthy farmers:

- It is this that reduces WAPDA's budget deficit because, under the provisions of the FRP, tubewell users are not supposed to pay full charges for electricity consumed.
- It is one of the reasons that the farmers are disinterested in adopting HEIS such as drip and sprinkler technologies. In fact, these technologies can provide services only under unabated power supplies.
- In the context of the flat rate policy, farmers never need to provide on/off switches in their tubewell installations and continue water extraction unabated; except when load-shedding. In this sense, electricity load-shedding is a blessing in disguise, as it helps in the conservation of the area's groundwater resources.

### *6.1.4 Degradation of Groundwater Quality*

To evaluate water quality, $_{p}$H value and salinity ratio are the two significant parameters. The former represents acidity/alkalinity of water, and salinity is a term used to describe the amount of salt in water (usually referred to in terms of total dissolved solids [TDS]). The World Health Organization (WHO) recommends water with a TDS concentration in the range of $\leq$1,000 mg/L as safe for human consumption. Water with a TDS ratio greater than this is considered saline.

**Table 6.3** Percentage frequency of tubewells by total bore depth (ft)

| | Depth | | | | | | | | | | | | | |
|---|---|---|---|---|---|---|---|---|---|---|---|---|---|---|
| Time | 50–100 | 101–150 | 151–200 | 201–250 | 251–300 | 301–350 | 351–400 | 401–450 | 451–500 | 501–550 | 551–600 | 601–700 | 70–800 | >800 |
| 1981–90 | 28.5 | 10 | 17 | 9 | 10 | 5.7 | 15 | 1.3 | 1.3 | 1.3 | 0.6 | – | – | – |
| 1991–2000 | 3 | 17.6 | 6.7 | 3.6 | 19.4 | 18 | 23.6 | 2.4 | 0.6 | 1 | 3.6 | – | – | – |
| 2001–05 | – | – | – | – | 3 | 2 | 14 | 8.5 | 17.7 | 5.5 | 27 | 16 | 5.5 | 1 |
| 2006–08 | – | – | – | – | – | 0.7 | 2.6 | 7 | 14.4 | 8.5 | 13 | 31.4 | 19.6 | 2.6 |

Source: Field Survey

**Table 6.4** Percent ratio of households by ownership of a quantity of dried out tubewells

| | Qty | | | | | | | | | | | | | | |
|---|---|---|---|---|---|---|---|---|---|---|---|---|---|---|---|
| Time | 0 | 1 | 2 | 3 | 4 | 5 | 6 | 7 | 8 | 9 | 10 | 15 | 24 | 25 | 30 |
| 1981–90 | 85.4 | 11 | 1.7 | 1 | 0.6 | – | – | – | – | – | – | – | – | – | – |
| 1991–2000 | 66.3 | 24.7 | 5.1 | 2 | 1.7 | – | – | – | – | – | – | – | – | – | – |
| 2001–05 | 34.3 | 31 | 22.5 | 6 | 2 | 3.4 | – | – | – | – | – | – | – | – | – |
| 2006–08 | 53.4 | 21.3 | 15 | 5 | 0.6 | 2.8 | 0.6 | 0.6 | – | – | – | – | – | 0.6 | – |

Source: Field Survey

**Table 6.5** Percent frequency of households with a quantity of tubewells

| | Qty. | | | | | | | | | | | | | | |
|---|---|---|---|---|---|---|---|---|---|---|---|---|---|---|---|
| Time | 0 | 1 | 2 | 3 | 4 | 5 | 6 | 7 | 8 | 9 | 10 | 15 | 24 | 25 | 30 |
| 1981–90 | 14.6 | 67 | 14.6 | 3 | 0.6 | 0.6 | – | – | – | – | – | – | – | – | – |
| 1991–2000 | 6.7 | 55 | 29 | 3 | 3 | 2 | – | – | – | – | 0.6 | 0.6 | – | – | – |
| 2001–05 | 7.3 | 22 | 36.5 | 12 | 12 | 4.5 | 2 | 0.6 | 1.7 | – | – | 1 | 0.6 | – | – |
| 2006–08 | 17 | 28 | 26 | 10.7 | 9 | 4 | 3 | 0.6 | – | 1 | – | – | – | 0.6 | 0.6 |

Source: Field Survey

**Table 6.6** Physico-chemical analysis of groundwater, February 2007

| Union council | Batezai | Alizai | Gangalzai | Manzaki | Manzari | Saranan | Yaro | Valley average |
|---|---|---|---|---|---|---|---|---|
| pH (6.5–8.5, WHO) | 8.75 | 8.53 | 8.5 | 8.5 | 8.33 | 7.96 | 8.51 | 8.44 |
| TDS (mg/l) (1,000, WHO) | 516 | 245 | 221 | 330 | 658 | 702 | 598 | 467 |

Source: Pakistan Council of Research in Water Resources, Quetta (unpublished)

To this end, we collected primary social data from all the sampled UCs/PCs, but secondary instrumental data for only 7 of 17 UCs/PCs. Results of the secondary data are given in Table 6.6. The $_p$H value of groundwater in the valley as a whole is within the normal range recommended by WHO; this value does not appear to limit its human and economic use. Nevertheless, in the case of Batezai, Alizai, and Yaro UCs, the normal range is slightly exceeded. Since a $_p$H value greater than 7 indicates alkalinity and tends to affect taste of water, it may become a problem in the near future, provided the current rates of the watertable fall are not checked.

Concentration of TDS in the year 2007 varies between the minimum value of 221 and 702 mg/L, with an average of 467 mg/L, well below the recommended level (Table 6.6). The higher TDS concentrations are found in Manzari and Saranan areas; the former is located in the northeastern piedmont zone of the valley, while the latter is located in the valley's central parts. It is interesting to note that the Alizai area recorded the second lowest TDS value despite that it is contiguous to the Manzari area.

Water quality perceptions were recorded through field survey in terms of simple arbitrary parameters (Table 6.7). It was concluded that during the period of 1981–90,

**Table 6.7** Percent ratio of respondents expressing various water quality values

| Year | For drinking purpose | | | For laundry use | | | For irrigation use | | |
|---|---|---|---|---|---|---|---|---|---|
| | Good | Fair | Poor | Good | Fair | Poor | Good | Fair | Poor |
| 1981–90 | 72.3 | 12 | 15.7 | 85.6 | 13 | 1.3 | 85.5 | 13 | 1.3 |
| 1991–2000 | 90 | 9 | 1 | 88.7 | 10 | 1 | 88 | 11 | 1 |
| 2001–05 | 89 | 9.7 | 1 | 87.4 | 11.4 | 1 | 88 | 11 | 1 |
| 2006–08 | 89 | 9.7 | 1.3 | 87.4 | 11.3 | 1.3 | 88 | 11.4 | 1.3 |

Source: Field Survey

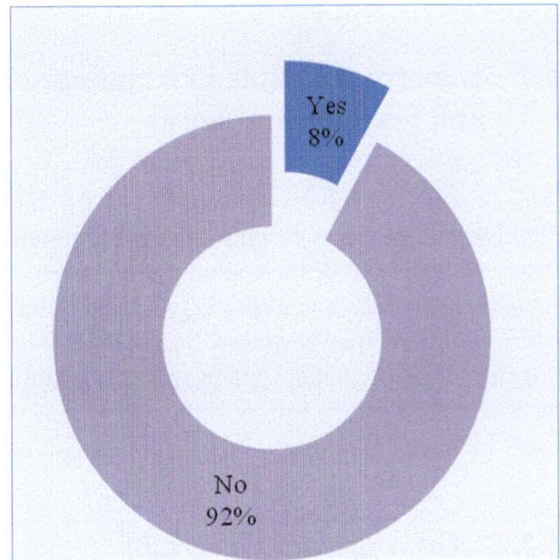


**Fig. 6.1** Percentage frequency of the farmers reporting salt color on soil top (Source: Field Survey)

the water quality was slightly brackish, which improved during 1991–2000, allegedly because of deeper drilling. But onward, deterioration is reported again for its domestic use (direct human use and laundry); for irrigation purposes, water quality has been reported as good to fair ever since.

When saline water is applied to agricultural fields over a long time, it paints them with a white color because salt particles do not evaporate with water. This can be used as an easily recognizable indicator of water quality. On average, 92 % of respondents reported no appearance of salt color in the soil (Fig. 6.1). Farmers who reported color were concentrated in five UCs/PCs; namely, Yaro (33 % of farmers), Batezai (12.5 %), Saranan (85.7 %), N. Malezai (20 %), and Segi (5.3 %) (Fig. 6.2). The logic is that these UCs/PCs are located in the central plain portion of the valley. While color is washed down by rainwater from the surrounding piedmont slopes, it remains deposited over the plain surfaces.

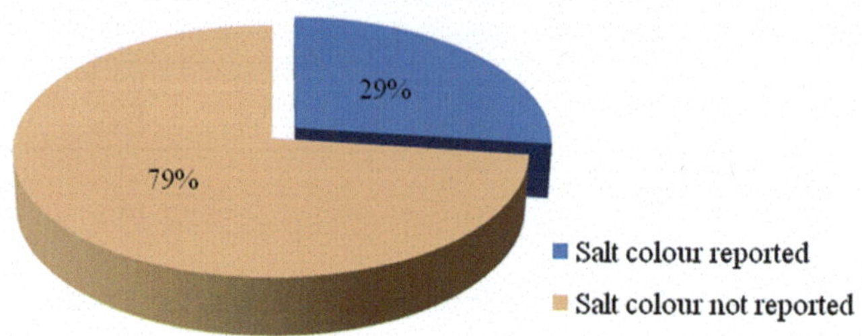


**Fig. 6.2** Percent ratio of the UCs/PCs showing salt color on soil top (Source: Field Survey)

## 6.2 Government Role in Aqua-agro Issues and Societal Response

Governments establish institutions, which have two basic elements: hardware deployed in the form of buildings and personnel; and legislation. In the case of Pishin Valley, both of these public prerogatives appeared weak. Both survey questionnaires (farmers and experts) indicated an institutional aspect of the problem to address certain touchy issues, including tubewell installation control, irrigation reforms, cropping pattern management, agriculture support policies, extension services, etc. The results follow:

### *6.2.1 Groundwater Governance*

The field survey asked farmers, "Does government control tubewell installation by any means (Appendix F, s. no. 6)?"; all 178 sample elements replied "No". Alhough the questionnaire contained queries about tubewell permits/NOC in case Government control of any type was reported, those queries were irrelevant for all respondents. In fact, rules exist on paper to have NOC from the District Water Committee before installing a tubewell, but in practice, no-one cares. Sure, the farmers are happy with not being checked by Government; however, the experts strongly advocate that groundwater abstraction be controlled through a licensing mechanism (Table 6.13); and that the license/permit be issued depending on the farm size and distance from pre-existing neighboring tubewells. Whereas the Groundwater Use laws of the GoB state that in plain areas with silty soil (Pishin Valley qualifies this criterion), the minimum distance between two nearest yielding points (karez, well, tubewell) must not be less than 750 ft; however, in practice, more than 90 % of the tubewells are in clear contradiction of those regulations (Table 6.8).

**Table 6.8** Distance between the nearest neighboring tubewells, 2008

| Distance (ft) | ≤100 | 101–200 | 201–300 | 301–400 | 401–500 | 501–600 | 601–700 | 701–800 | 801–900 | 901–1,000 | ≥1,001 |
|---|---|---|---|---|---|---|---|---|---|---|---|
| % frequency of tubewells | 4 | 39.6 | 24.7 | 7 | 9.7 | 2.6 | 4 | 0.6 | 0.6 | 0.6 | 6.5 |

Source: Field Survey

## 6.2.2 *Groundwater Conservation*

The concept of groundwater conservation means two things:

- Economization in water use; i.e., increase in water efficiency per unit of its use.
- Physical enhancement of groundwater reserves; i.e., replenishment of the aquifer.

This study examined the public policies/strategies that are someway directed toward both these dimensions of the local groundwater resources.

### On-Farm Water Management

This means field-level water management. This aspect was analysed through the parameter of lining/piping of irrigation infrastructure, including water tanks and water delivery courses. The study found that only 20 % of the irrigation tanks are lined with concrete (Table 6.9). The tanks are, in fact, filled with water during the night through tubewells, and the water is applied during the day to crops by flood method. During the day the tubewells are also operated to continue contribution to the tanks but the night-time storing contributes to rapid irrigation of the farms, in view of the small discharge capacities of tubewell and prolonged load-shedding during the day. Likewise, lining and piping ratio of the watercourses was also found to be nominal (Table 6.9).

Of the total number of tubewell irrigators in 2008, 11.3 % had some of the water-courses lined, while 5 % had some piped (Table 6.9). Of the 11.3 % farmers with lined watercourses, the majority (44.4 %) had them between 41 % and 50 % of the total length of their watercourses. This group is followed in terms of frequency by 33.3 % of the farmers with a ratio of 21–30 % of their watercourses lined. On the same pattern, the modal values in the case of piped watercourses are ≤10 % and 81–90 % for 25 % of the farmers each.

Lining (cementing) and piping (PVC) of irrigation infrastructure can serve the water conservation goal in several ways. The low ratio in this regard is indicative of the wastage of water in irrigation works, which reduces water productivity per unit of use. Total liners/pipers accounted for 48 of 159 irrigation farmers in the year 2000. Except for one farmer, who had met the entire cost of lining/piping himself, the rest were funded by Government by various ratios of their actual expenditure; in the majority of cases by 60–70 % of the total cost (Table 6.10).

**Table 6.9** Percentage ratio of lined/piped irrigation tanks and courses, 2008

| | Lined irrigation tanks (%) | Percentage of farmers per a percentage of lined watercourses | | | | | | Percentage of farmers per a percentage of piped irrigation courses | | | | | | |
|---|---|---|---|---|---|---|---|---|---|---|---|---|---|---|
| | | Group farmers | % of lined water courses | | | | | Group farmers | % of piped water courses | | | | | |
| | | | ≤10 | 21–30 | 31–40 | 41–50 | 61–70 | | ≤10 | 11–20 | 21–30 | 51–60 | 71–80 | 81–90 |
| Total irrigators 159 (89.3 % of the sample) | 20 | 11.3 | 11 | 33.3 | 5.6 | 44.4 | 5.6 | 5 | 25 | 12.5 | 12.5 | 12.5 | 12.5 | 25 |

Source: Field Survey

**Table 6.10** Govt. financial share in lining/piping of tanks/water courses

| % of total cost | 0 % | 60–70 % | 71–80 % | 81–90 % |
|---|---|---|---|---|
| Frequency of beneficiaries | 1 | 25 | 12 | 10 |

Source: Field Survey

It is encouraging to see Government support in water conservation; however, given the overwhelming poverty in the area, the funding level may be raised indiscriminately for all farmers. The PVC piping method appears to be more promising than lining, due to its special usefulness in curtailing evaporation loss.

### Introduction of Micro-irrigation Technology

In arid climates, to conserve water and increase its productivity per unit of use, micro-irrigation technology is the most recommended modern option. This technology has several forms, but drip (trickle) and bubbler systems are considered more appropriate for local needs. Since the general climatic and soil characteristics of Pishin Valley are congenial for growing fruits if irrigation water is available, introduction of micro-irrigation technology is imperative to enhance water application efficiencies. These systems are very apt for tree crops such as the fruit orchards in Pishin Valley. The drip system applies small volumes of water right to the base of the plant where the roots are concentrated. The bubbler system applies water to crops through tubular-shaped bubblers attached to a pipeline. Like the drip system, this technology is also suitable for fruit trees, particularly those above 2 years of age. A third type of efficient irrigation systems is the sprinkler, which is recommended for field crops. Sprinklers create rainfall-like conditions by moving around and spraying water over crops within a set range of distance. The drip technology was installed for the first time at Quetta District of Balochistan during the early 1980s on a limited scale with the financial support of the Agricultural Development Bank of Pakistan. In subsequent years, it disseminated to some other parts of the province, under some development projects funded by international, national, and provincial agencies. However, these systems could not gain enough popularity among the farmers for reasons including the high capital cost, lack of awareness and consciousness toward the emerging water crisis, and cultural inertia, etc.

The field survey asked about the use or otherwise of any type of HEIS (Appendix F), but the query received a "No" reply from 100 % of respondents. The survey further asked for reason(s), the results of which are depicted in Fig. 6.3. Lack of awareness was the leading cause for not using any HEIS. The next major factor was the prolonged load-shedding, as these systems are motor operated; other factors included low harvest, high capital cost, and the controversial flat rate policy, etc.

The fact that the efficient micro-irrigation technology has no presence in the area is a failure of the Government, particularly when respondents claim no awareness of

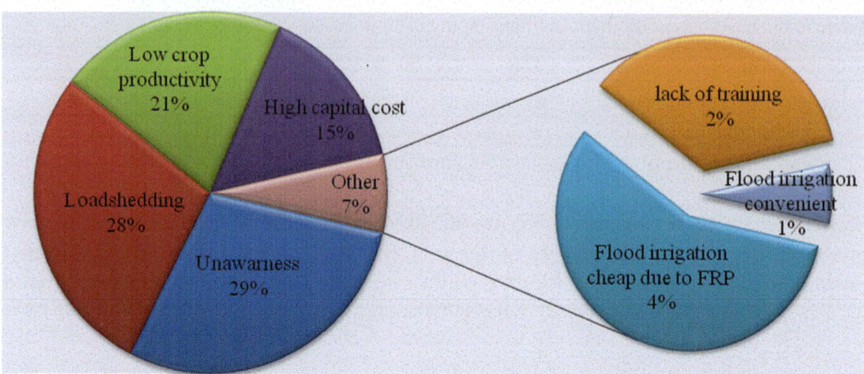


**Fig. 6.3** Reasons against adoption of micro-irrigation systems (Source: Field Survey)

or training in these systems. Among the suggestions/demands by farmers, provision of HEIS for free or at subsidized rates is very frequent. Likewise, the experts also consider subsidized provision of HEIS to farmers as one of the best choices for sustainable progress of tubewell farming in the area (Table 6.13).

## Rainwater Harvesting

Rainwater harvesting is not a modern phenomenon; it is among the oldest irrigation technologies in human history, but its forms may vary. The concept of rainwater harvesting for agricultural purposes includes run-off storage and its direct use through diversion structures. In the area under study, rainwater storage structures are referred to as 'delay action dams' (DADs). The DADs are constructed on the bottle necks of inundation channels (hill torrents) within the hilly terrain just close to plain agricultural areas.

The concept of DADs was first picked up in Balochistan's uplands in the 1960s and strengthened in subsequent eras. Its popularity skyrocketed during the 1980s and 1990s as an urgent remedy for the rapid groundwater depletion and the poor recharge rate in the region. However, in later years, their usefulness was questioned in the context of improper biological watershed management and consequent large-scale siltation phenomenon in these reservoirs. Although the DADs culture prevails, the heavy investment allocation for them is a controversial issue between those with sincere expertise and the bureaucratic and political elite. A leading section of experts subscribe to the thought that in the overall ecological setup of the Balochistan uplands, which includes the study area, the DADs are arresting rainwater just to subscribe to atmospheric moisture (evaporation) rather than to the aquifer at most.

To serve the Pishin sub-basin's aquifer (including the valley in focus), some 14 DADs projects were completed until 2008 (Map 3.9), and several others were either

**Table 6.11** Nearest DAD's distance from command area and its contribution to aquifer, 2008

| Nearest DADs distance from command (km) | | | | Contribution to aquifer recharge | | If yes, to: | | |
|---|---|---|---|---|---|---|---|---|
| ≤5 | 6–10 | 11–15 | ≥16 | Yes | No | Karez | Tubewell | Both |
| 51 | 44 | 25 | 3 | 2 | 131 | 2 | 0 | 0 |

Source: Field Survey

under construction or in the pipeline. Our field survey discussed their status with the farmers and a sample of relevant expert officials. Of the 133 answering farmers, the closest DAD to many of them was no further than 5 km from their farms (Table 6.11). However, only two reported their significant contribution in the watertable recharge, which was observed by them only in karezes, not in tubewells. Karezes are artificial shallow water-yielding systems emerging from the foothills close to the DADs. The experts also suggest only a marginal effectiveness of DADs in aquifer recharge (Table 6.19). In the suggestions list from both the experts and the farmers, construction of effective recharge structures, side by side with protection of recharge areas by proper watershed management, appeared frequently (Tables 6.13 and 6.17).

### 6.2.3 *Crop Management and Extension Training*

A view states, "when you can't change the world, change yourself." When irrigation opportunities have shrunk, and no promising remedy has been found, let the dependent cropping pattern be adapted to ground realities. In this regard, low-delta/high-value and drought-tolerant/resistant crop varieties can make a change in the poverty line of the area and the whole nation at large. Our survey asked the farmers whether or not the Government checks their crop selection? All reported "No". For a possible "Yes" answer, the questionnaire contained queries regarding the restricted and/or supported crops, but these sections remained blank. Similarly, 100 % of the farmers indicated that they had neither been invited on any extension training workshop from public sector departments, nor attended any such training from any other source.

In fact, the Government is not interfering in the cropping pattern, nor has it provided any extension training to that end, yet farmers, based on their own knowledge and skills have made some alterations in crop selection (Table 5.9). For example, horticulture, which has been the dominant stakeholder in the area's under-cultivation acreage, is leaving place to annual field-crops, which, if destroyed in any one unfavorable season, may not cause a long-term loss in a household's budget. That is, an orchard, once lost, requires several consecutive good years to reproduce; whereas small field crops, like wheat and vegetables, etc., are a season-to-season gain/loss business. That is the lesson the farmers of the area have learnt from the drought of the recent past (Appendix F). In lieu of tree fruits (e.g. apple, apricot, etc.), vegetables, tobacco, and wheat have been adopted; in some cases, grape orchards, which is a relatively low-delta fruit, have been adopted in lieu of apple orchards. Although vegetables and tobacco are high-delta crops, a little land devoted to them earns much because they are also high-value crops.

When the line experts were asked how water shortages have affected cropping pattern, they overwhelmingly highlighted the change in terms of the shift from horticulture to field crops (Table 6.14); followed by the shift from low-value to high-value crops. With relatively small significance level, as per the interview schedule's terminology, the line experts suggested a change from high-delta to low-delta crops. This means, despite not being checked by Government, water conservation is still not the concern of the people; rather, they are behind profit making at all costs. Of the experts, 13.3 % reported no significant change in cropping patterns, even under the current water scarcity scenario.

### *6.2.4 Implications of the Flat Rate Policy*

The flat rate policy, under which a fixed amount of money is paid by a tubewell owner, irrespective of the actual consumption of electricity, is not unique to Balochistan province in Pakistan. Rather, this policy was introduced nationwide during the 1970s in order to encourage groundwater-dependent farmers to expand tubewell farming and thus increase agricultural production. Indeed, this policy has been instrumental in the fast expansion of tubewell-irrigated agriculture in the Balochistan uplands, including the study area. Consequently, although tubewell farming increased many-fold, it simultaneously promoted wasteful use of the inherently scarce groundwater resources, thus challenging the sustainability of the area's agricultural economy. Transforming from a fixed to a consumption-proportionate tariff of electricity in irrigation tubewells is the obvious way to promote water conservation; however, the primary impediments to this are the lack of political will and institutional capacity to bring it to action. Reports tell that many influential farmers do not even pay the already low flat rate for electricity, which is only Rs. 4,000 (about $US42, June 2014) per month per tubewell.

The survey tool contained queries regarding how farmers value the flat rate policy and how they may react in case otherwise. They were asked, "Is it true that the FRP is encouraging water overuse?". All respondents replied "Not at all". Next, the survey asked, "Will you don't mind if the government calls-off the Policy?" All said, "No, we shall not allow it." The survey asked, "How shall you respond if it was called-off anyway?" A total of 97 % intended mass agitation against any such announcement; 2 % said, "It is ok, but the Government must provide us the HEIS free of cost"; 1 % asked for funds to line their irrigation infrastructure completely. We further probed the significance of the FRP to farmers by asking them what they may opt to do if they had to pay the full electricity tariff; 81.5 % said they shall give up cultivation anymore, 4 % replied they would use a diesel motor, 1.7 % would look at dry farming, and only 1 % resolved to continue with tubewell irrigation as usual. The survey also asked whether the current flat rate (rupees 4,000) was appropriate.

**Table 6.12** Survey B, question 1; Why is there so much cultivable-waste land in Pishin Valley?

| | | Votes frequency (modal values bold faced) | | | |
|---|---|---|---|---|---|
| S. no. | Possible answers | Untrue | Highly true | Fairly true | Slightly true |
| A | Lack of irrigation water | – | **14** | 1 | – |
| B | Abundance of land disputes | 3 | 2 | 2 | **8** |
| C | Lack of population/laborers | **6** | 1 | 5 | 3 |
| D | Lack of interest of owners | 3 | 4 | 1 | **7** |
| E | Poverty of owners | 2 | **8** | 3 | 2 |
| F | Abundance of state lands | **6** | 2 | 3 | 4 |
| G | Any other reason | Load-shedding = 2; Precipitation deficiency = 4 | | | |

Source: Field Survey

A total of 95 % replied positively and the rest said, "No, it too is high, because we hardly find electricity to use due to prolonged load-shedding." Nonetheless, the experts highly recommended calling off the FRP (Table 6.22) and they further highlighted that the main drivers of the FRP are the fear of public resentment and the WAPDA's inability to curtail electricity thefts and collect normal bills according to the actual meter reading (Table 6.23).

### 6.2.5 *Farmers' Suggestions/Demands*

The survey tool used open-ended questions to invite suggestions from the farmers; 19.6 % did not participate in those queries. A total of 162 suggestions were made (some gave one suggestion, others gave several). Of these, 45.6 % were in support of the construction of effective small dams (DADs), 17.3 % each for load-shedding elimination and funds for irrigation infrastructure lining. Of the farmers, 8.6 % asked for de-siltation of the BKK, which was an important source of surface water irrigation, as well as a means for aquifer recharge; 4.9 % demanded supply of HEIS free of cost; and 3.1 % suggested their subsidized supply. The remaining 3.6 % of the farmers wished for the Government to play a role in rehabilitation of their age-old traditional karez system.

### 6.2.6 *New Vision from Experts*

Our relevant survey tool offered a number of touchy issues to officials of the line departments for their deliberate opinions (Appendix G). The results are presented query by query in separate tables. On the issue of the presence of enormous cultivable-waste land in Pishin Valley, the unopposed and most significant factor responsible for that reality was indicated as being the acute shortage of irrigation water (Table 6.12); poverty of the landowners was also reported among the key reasons, but thin population and presence of public lands were largely rejected as significant.

**Table 6.13** Survey B, question 2; Suggest new strategies for sustainable farming in the Valley

| | | Votes frequency (modal values bold faced) | | | |
|---|---|---|---|---|---|
| S. no. | Possible answers | Untrue | Highly true | Fairly true | Slightly true |
| A | Groundwater abstraction control by tubewell licensing | 1 | **11** | 3 | – |
| B | Subsidized provision of HEISs | – | **12** | 2 | 1 |
| C | Emphasis on intercropping | 1 | **10** | 1 | 3 |
| D | Adoption of low-delta crops | 1 | **11** | 2 | 1 |
| E | Conservation of recharge zones in the watershed | – | **13** | 1 | 1 |
| F | Rehabilitation of traditional sailaba, and khushkaba farming | 2 | **6** | 5 | 2 |
| G | Any other suggestion | Forestation = 1; Effective recharge structures = 1 | | | |

Source: Field Survey

**Table 6.14** Survey B, question 3; How water shortage has affected the cropping pattern?

| | | Votes frequency (modal values bold faced) | | | |
|---|---|---|---|---|---|
| S. no. | Possible answers | Untrue | Highly true | Fairly true | Slightly true |
| A | Shift from orchards to field crops | 2 | **11** | 2 | – |
| B | Shift from low- to high-value crops | 2 | **10** | 3 | – |
| C | Shift from high- to low-delta crops | 2 | **8** | 5 | – |
| D | Any other change | No change = 2 | | | |

Source: Field Survey

As the way forward to avert the foreseeable collapse of tubewell farming in the region, the line experts confirmed the great relevance of all the remedies presented in Table 6.13, but the overwhelming among them was the watershed management and introduction of efficient micro-irrigation technologies. In addition, protection and enhancement of the area's natural green cover and construction of effective recharge structures (of the type of existing DADs or others) were also suggested.

On alterations of cropping patterns, although some denied any considerable change in this regard, many perceived a great change in all three possible areas of change as invoked under question 3 in the questionnaire B (Table 6.14).

Regarding the level of public interest in the cropping pattern issue, the survey results concluded that there is no restriction on any type of crop on the basis of its water use characteristics, but that low-delta crops are supported by many accounts, such as awareness generation, and seed and marketing arrangements, etc. (Table 6.15).

On the aquifer issues, the survey found that experts believed the root cause of the rapid and excessive watertable decline to be overuse of water in irrigation, followed by climate change in terms of precipitation deficiency, elimination of recharge areas, and immense expansion of tubewell farming (Table 6.16).

What should be done then? We asked this of the experts by offering four lines of possible action. They urged two lines in particular; i.e., to construct more effective

**Table 6.15** Survey B, question 4; What are Govt. strategies to popularize low-delta crops?

| | | Votes frequency (modal values bold faced) | | | |
|---|---|---|---|---|---|
| S. no. | Possible answers | Untrue | Highly true | Fairly true | Slightly true |
| A | Restricts cultivation of high-delta crops | **15** | – | – | – |
| B | Subsidizes seeds, etc. of low-delta crops | 2 | **5** | 5 | 3 |
| C | Educates farmers for low-delta crops through training and media | 4 | **7** | 1 | 3 |
| D | Increases support prices for low-delta crops | **6** | 5 | 1 | 3 |
| E | Arranges special marketing for low delta crops | 6 | **7** | 1 | 1 |
| F | Any other strategy | Nil | | | |

Source: Field Survey

**Table 6.16** Survey B, question 5; What are reasons of watertable depletion in Pishin Valley?

| | | Votes frequency (modal values bold faced) | | | |
|---|---|---|---|---|---|
| S. no. | Possible answers | Untrue | Highly true | Fairly true | Slightly true |
| A | Precipitation deficiency | – | **10** | 5 | |
| B | Recharge zones elimination | 3 | **9** | 3 | |
| C | Immense tubewell farming | 2 | **9** | 4 | |
| D | Water over-use in irrigation | – | **14** | 1 | |
| E | Any other reason | Nil | | | |

Source: Field Survey

**Table 6.17** Survey B, question 6; Suggest new strategies to avert watertable depletion

| | | Votes frequency (modal values bold faced) | | | |
|---|---|---|---|---|---|
| S. no. | Possible answers | Untrue | Highly true | Fairly true | Slightly true |
| A | Shift from agriculture to industry or else less water-intensive economy | 2 | 5 | **8** | – |
| B | Construction of effective recharge structures | – | **13** | 2 | – |
| C | Use of HEISs | – | **13** | 2 | – |
| D | Ban on further irrigation tubewells | 5 | **6** | 3 | – |
| E | Any other suggestion | Extension training= 1; Adoption of drought tolerant crops= 1 | | | |

Source: Field Survey

recharge structures (like infiltration galleries) and widespread introduction of the HEISs. Besides those lines of action, shifts from agricultural to industrial or other less water-intensive economies was also supported as a wise option. Further, a complete ban on installation of further irrigation tubewells was also envisioned by a good number of the experts (Table 6.17).

On the issue of groundwater governance, the experts gave a greater confirmation to the lack of political will as being the main hurdle in enforcement of

**Table 6.18** Survey B, question 7; What impedes enforcement of groundwater regulations?

| | | Votes frequency (modal values bold faced) | | | |
|---|---|---|---|---|---|
| S. no. | Possible answers | Untrue | Highly true | Fairly true | Slightly true |
| A | Lack of political will | – | **12** | 3 | – |
| B | Poor accessibility of the area | 3 | **7** | 5 | – |
| C | Institutional weaknesses enforcing agencies | – | **9** | 6 | – |
| D | Institutional corruption | 1 | **11** | 3 | – |
| E | Weak regulations | 2 | **8** | 5 | – |
| F | Any other reason | New regulations = 1; Sensitization of institutions = 1 | | | |

Source: Field Survey

**Table 6.19** Survey B, question 8; In which use DADs have been successful?

| | | Votes frequency (modal values bold faced) | | | |
|---|---|---|---|---|---|
| S. no | Possible answers | Untrue | Highly true | Fairly true | Slightly true |
| A | Quite effective in long-term aquifer recharge | 3 | **5** | 3 | 4 |
| B | Quite effective in short-term aquifer recharge | 1 | **7** | 4 | 3 |
| C | Marginally effective in aquifer recharge | 1 | **10** | 3 | 1 |
| D | Effective in flood control | – | **11** | 3 | 1 |
| E | Useful as sources of pot water/stock water/picnic sites | 2 | **11** | 1 | 1 |
| F | Any other benefits | Nil | | | |

Source: Field Survey

groundwater regulations (Table 6.18). The other hurdles in this connection were reported to be institutional corruption and weaknesses, and the regulations themselves being weak, etc.

The DADs, which were constructed in the Balochistan uplands with a view to enhance the potential of the aquifers, found little favour with the experts regarding their effectiveness in the key purpose for which they were deemed (Table 6.19). The dams have proved to be only marginally effective in aquifer recharge, but as saviors from floods, their utility is acknowledged a great deal (11 votes in the 'highly true' category). In uses other than aquifer recharge, the DADs were thought to be a good means of pot and livestock water and as picnic sites for local inhabitants.

According to the experts, the failure of the DADs has been common but not universal; in some cases they have really served the purpose of aquifer recharge. However, the two main causes of their failure were reported as:

1. Choking up of the subsurface infiltration pores.
2. The thick layers of sediments brought in by hill torrents from the barren watershed (Table 6.20).

What or who is the stumbling block in adoption of the very useful micro-irrigation systems? The top decisive factor was revealed to be the lack of awareness

**Table 6.20** Survey B, question 9; In failure cases of DADs what was/were the reason(s)?

| | | Votes frequency (modal values bold faced) | | | |
|---|---|---|---|---|---|
| S. no. | Possible answers | Untrue | Highly true | Fairly true | Slightly true |
| A | Quick sedimentation, thus short storage life | 2 | **13** | – | – |
| B | Choking of subsurface pores quickly, and thus no contribution to aquifer recharge | 1 | **14** | – | – |
| C | Any other shortfall | Site selection accuracy might have failed = 1 | | | |

Source: Field Survey

**Table 6.21** Survey B, question 10; What hurdles the use of HEISs? (Appendix G)

| | | Frequency (modal values bold faced) | | | |
|---|---|---|---|---|---|
| S. no. | Possible answers | Untrue | Highly true | Fairly true | Slightly true |
| A | High O & M cost | 1 | **9** | 5 | – |
| B | Lower production than flood irrigation | 4 | 2 | 3 | **6** |
| C | Poor availability of equipments and spare parts | 2 | **7** | 4 | 2 |
| D | Unawareness for use of water conservation | – | **12** | 3 | 1 |
| E | Unawareness about HEIS | 1 | **12** | – | 2 |
| F | Cheaper water pumping for flood irrigation due to FRP | 2 | **9** | 2 | 2 |
| G | Any other hurdle | All the above factors are contributory = 1 | | | |

Source: Field Survey

**Table 6.22** Survey B, question 11; Suggest new strategies to make HEIS popular

| | | Votes frequency (modal values bold faced) | | | |
|---|---|---|---|---|---|
| S. no. | Possible answers | Untrue | Highly true | Fairly true | Slightly true |
| A | Subsidized supply of HEIS to farmers | – | **14** | 1 | – |
| B | Flat rates policy be called off | 3 | **9** | 3 | – |
| C | Use of HEIS be made a condition for power subsidy | 3 | **9** | 3 | – |
| G | Any other suggestion | Public demonstration of HEIS by line agencies = 2 | | | |

Source: Field Survey

or consciousness of the people about how crucial water conservation is to their sustainable sustenance and how effective the HEISs are in this regard (Table 6.21). High capital and maintenance costs for HEISs and the flat rate policy were also blamed as the latter ensures availability of relatively cheaper, though wasteful, water for irrigation.

The new or revised vision that the experts proponed in support of HEISs was their subsidized supply to farmers as a top priority because the farmers are generally poor and are fearful about their failure and the consequent loss of money (Table 6.22). If they could obtain the technology at quite nominal rates, they would agree to take the risk of their failure. Simultaneously, the FRP may be called off altogether, or at least be made conditional with the use of HEISs by the beneficiaries.

**Table 6.23** Survey B, question 14; Which motives drive the FRP?

| | | Votes frequency (modal values bold faced) | | | |
|---|---|---|---|---|---|
| S. no. | Possible answers | Untrue | Highly true | Fairly true | Slightly true |
| A | Vested interests of decision makers | 1 | **8** | 3 | 3 |
| C | Fear of public resentment if FRP called off | 2 | **10** | 1 | 2 |
| D | WAPDA's inability to check electricity thefts and recover full tariff | – | **13** | 2 | – |
| E | Farmers' welfare, since they cannot urn due to huge irrigation cost | 2 | **9** | 4 | – |
| G | Any other motive | Nil | | | |

Source: Field Survey

Under general questions, 93 % of the officials blamed the FRP as the cause of over-abstraction and misuse of water. A total of 73 % suggested lifting the FRP immediately; 7 % said, "not without other necessary arrangements"; and the other 20 % hesitated to say something about it. Those who said "No" justified it with the high normal electricity tariff and a possible huge loss in net farm income. In terms of deterrents that have been preventing the Government from withdrawing the FRP, the WAPDA's inability to check theft and regulate proper metering was suggested as the main factor (Table 6.23). The risk of political loss to any sitting government as a result of public resentment also has a role in this situation, along with other minor considerations.

Among the line expertise, one-third expected serious public resentment if the FRP were to be dropped and the others did not expect it to be worrisome. It was pointed out that if complete withdrawal of the FRP was not feasible, let the tariff share by farmers and government be 50:50. Likewise, subsidy in tariff up to a certain upper limit of meter reading, generation of mass awareness toward resource conservation, and poverty alleviation by expanding job opportunities were seen as viable steps to convince the public to accept the dropping of the FRP peacefully.

In addition, extension trainings, cropping pattern check, dissemination of drought-tolerant/resistant crop varieties, media awareness campaigns, rehabilitation of the traditional karez systems, elimination of load-shedding, area-targeted research, etc. appeared as lines of vision of the experts.

## 6.3 Summary

In this age of climatic disorder, when the normal regimes of temperature and precipitation of a region usually appear now as alternating catastrophic episodes of drought and floods, agriculture is not viable without a reliable means of irrigation. With the same logic, the historically meager rain-fed lands, irrespective of their size and location, have now been mostly deserted because the input cost and the required net income margin to live upon them have increased many-fold. Obviously, under the mentioned scenario of climate change, the most trustworthy source of irrigation can be the groundwater. Whether due to anthropogenic shortcomings or exclusively because of natural limitations, the high dependency on groundwater irrigation has greatly exhausted the potential of the aquifers in many dry areas of the world, including Balochistan province at large and the case study area in particular.

This study found that aquifer depletion has been a regular phenomenon in Pishin Valley, but its magnitude varied from area to area depending upon the topography, precipitation fluctuations, and the magnitude of pumping for irrigation purposes. The farmers are now pumping beyond the depth of 800 ft, with an overall high irrigation cost. Rather, the experts suggest that this is the level where use of the term "aquifer mining" is justified, which means an incurable destruction of an aquifer. Although there is presently no significant groundwater quality problem, it may become an issue as aquifer mining and other anthropogenic pollution activities increase. In addition to the appearance of salt coloring on the soil surface, the increasing use of plant drugs and chemical fertilizers in agricultural lands is also polluting the aquifer. Moreover, this chapter provided important facts and discussion concerning the development and effectiveness of agricultural and water use strategies.

# Chapter 7
# Conclusion, Key Findings, and Suggestions

**Abstract** Research is the means to problem solution. The research at hand investigated the interdependent aspects of the two-pronged agro–irrigation problem of the Pishin Valley as a representative sample part of the grand phenomenon taking place in the whole of Balochistan province, particularly its uplands. This chapter integrates the results of the research to a meaningful conclusion, which is to be pursued further in the course of time. It recapitulates on the key contextual aspects of the study, discusses the salient features of research methodology and the data used, articulates the key findings understood thereby and, where anthropogenic weaknesses and shortcomings are involved in the problem, suggests viable corrective measures accordingly.

**Keywords** Pishin valley • Irrigated agriculture • Conclusion • Findings • Suggestions

## 7.1 Conclusion

In the present era of climate change, irrigation farming in many parts of the world has begun yielding diminishing returns. The Balochistan province in general, especially its plateau or highland zone, also subscribes to that phenomenon and is inherently more vulnerable due to its small resource base. Agriculture is one of the most important sectors of the province's economy, and contributes around 52 % of its GDP and employs about 65 % of its labor force. This sector is the major user of land and water in the province; more than 1.93 million hectares are under cultivation and about 97.5 % of all available water in the province is used in agricultural production. The province has become one of the greatest water-poor regions of Pakistan. In most areas here, farmers are pumping groundwater faster than it is being replenished by nature, causing a consistent drop in the watertable. Groundwater over-abstraction, in the sense of greater discharge than recharge, is now the single biggest challenge to Balochistan's irrigated agriculture. However, the surface

A.S. Khattak, *Mutual Sustainability of Tubewell Farming and Aquifers: Perspectives from Balochistan, Pakistan*, Advances in Asian Human-Environmental Research, DOI 10.1007/978-3-319-02804-0_7, 

water resources, though inherently scarce, if used judiciously and conserved, can ensure the sustainability of irrigation agriculture for the next few generations.

This research was conducted within the physical, socio-economic, and policy framework of the Pishin Valley as a representative case for Balochistan province at large. We gave enough space to description of elements of the local environment, mostly through secondary information, and found all of them having significant imprints, implicitly or explicitly, in the aqua-agro past and present scenarios of the region. For instance, we found that relief was a factor in the DTW as, in the piedmont zone, the watertable was always (1981–90, 1991–2000, etc.) reported as deeper than in the valley's low-lying central floor. Likewise, it surfaced that the watertable depletion rate per annum was increased by the drought of 1998–2004, which is validation of the role of climate in the discharge and recharge potentials of the aquifer. The role of soil quality was illuminated by satellite data as well as primary data of agricultural land-use, which established that the better quality piedmont soils were densely cultivated, despite the watertable being deeper there, compared with the low-quality saline soils of the valley's flat center. By the same pattern, the influence of other local characteristics, such as population density, education, farm size, and tenancy culture, etc., were found to have a role in setting trends in tubewell farming and aquifer potential.

Farmers are the key stakeholders in the problem studied. Therefore, understanding their socio-economic profile was imperative for wider applicability of the results. We found a joint-family system in the area, with an average household size of around 32 individuals. The joint-family system may have some merits, but its demerits in agricultural occupation cannot be ignored as it is, by nature, a laborious job. For instance, the common property tragedy was obvious in the society in that an average of only 3.6 individuals per household were being spared to do all work on 18 acres of irrigation farms per head. Considering the labor intensiveness of irrigation farming in general and of horticulture in particular, we believe that this size of labor input is insufficient. In fruit orchards, agricultural machinery (e.g. tractors) cannot operate; rather, weeding, heaving, etc. are all done manually with spades several times a year. Maintaining tubewell machinery is another troublesome job. It was also found that one-third (32 %) of the farmers were illiterate and 60 % of the rest had quit education at matriculation level. This means that many of the farmers could not study agriculture-related literature or understand the Urdu language properly (this is Pakistan's national language and is the main medium of the country's media). Such deficiencies in the farming population contribute to the dwindling contribution of farms in the households' annual budgets. While 46.6 % of households were making their complete living through farming in 1981–90, this ratio steadily dropped to 37 % in 2005–08. Consequently, we noted that about 69 % of the originally farming households had started side jobs to fill the gap in farm income. Although we deliberately interviewed aged farmers, because only they could answer questions relating to something far in the past as well as having field knowledge, we understood that farming is practiced mostly by upper middle-age people and the youth are less interested.

The study picked up several objectives to meet. The first was to assess expansion and intensification in agricultural land use from 1981 to 2008 through multiple data sources. The analysis concluded that land under cultivation increased from the 1981 level, with more than 4 % growth rate per 10 years until 1990 and this was the first and last boom period in the 28-year history for which this study accounted. The bust that started during 1991–2000 has been proceeding steadily unabated but the per annum rate varies. During 2001–04, it was 0.22 %, while during 2005–08, it decreased to 0.1 % annually. Thus, after 2004, we see some stabilization in the level of acreage under cultivation, which may be due to the revival of wet periods after the prolonged drought of 1998–2004. The relatively sharp decline of under-cultivation percentage coincides with the peak period of the aforementioned drought.

The analysis further explored that in the total land under cultivation, the irrigation farm size steadily increased unabated from the start of the study period until 2004. However, since then, a halt in the growth, rather a nominal backdrop at the rate of 0.01 acres per annum prevailed. Whatever development was taking place in irrigation farming, it was primarily because of expansion of irrigation – exclusively tubewell irrigation – to new lands. The slight downward trend beyond 2004 can be quantitatively attributed to reduction in aquifer potential, as the data suggest. As a whole, although irrigated acreage has consistently increased, its pace has not been as fast as it should be given the population expansion and availability of plenty of cultivable-waste lands. The prime deterrence appears to be aquifer limitations and accordingly the high cost of irrigation. It is clear that, day after day, the economy of the area is transforming from agriculture to service sector employment.

Moreover, an interesting feature is that, while new land is being brought under plough by tapping new points in the overall aquifer, some previously cultivated lands are subscribing to the process of desertification for many reasons but primarily due to watertable drawdown and diminishing financial returns. The desertification process originated in the 1990s and peaked during 2001–04 when a sum of 431 acres of land was reported deserted within all sampled respondents. However, not all of the desertification is a permanent feature, in some cases the area once lost is reclaimed whenever the owners become able to arrange a means of irrigation – obviously, a tubewell.

Green cover in the valley was delineated from five Landsat imageries of 30 m resolution each via supervised classification method. Since the area has no conspicuous natural greenery, whatever greenery appeared on the images was presumed to be agriculture, particularly of the irrigated type. The images presented kharif (summer) cropping with intervals of nearly 5 years. The cultivated area trended almost the same growth/fall patterns as was revealed by the field survey (questionnaire) data. But here, instead of durational, the data was point-data of single seasons (kharif) with intervals. The magnitude of cultivation was 4.7 % of the total area of the valley in 1989, which quickly increased to 8.2 % in 1991, just 2 years later. This means the growth rate was 1.8 % per year meanwhile. The trend also increased in 1996 but with only 1.6 % total difference in a time interval of 5 years. A huge slump in cultivated acreage occurred in the year 2000, illustrating the effect of the drought in 1998–2004. The year 2005 recaptured the usual precipitation regime, hence it is the most green.

In the individual UCs/PCs, the year 2000 was found to be the least cultivated everywhere in the valley. All of them show rising trends of green cover from 1989 to 1996. Although 2005 is greener than 1996 as a whole, some UCs/PCs are found to be more cultivated in 1996 than in 2005; for instance, Q. Abdullah, D. Khanzai, Alizai, and Torashah. The reason for this is the lower depletion of the watertable there on account of being located in the northern piedmont zone of the valley, where recharge potential is high due to several check dams in the neighborhood.

Regarding farm intensity, we found farm-fallowing cases with a ratio of 32 % in the kharif season and 27.6 % in the rabi season out of the total respondents. Stated another way, 68 % of farmers in were cultivating farms with 100 % intensity in the kharif season and 71.4 % in the rabi season. The kharif season farm fallowing was forced mainly by water scarcity and that in the rabi season was mainly due to frost. However, the portion of land fallowed in both seasons was not greater than 20 % of the under-cultivation land. We found that farm fallowing came to practice in 1975 and the number of farmers in this culture increased until the year 2000; stabilized until 2005, and re-emerged with a ratio of 5 % during 2006–08. Among the reasons for farm fallowing, the highest was water scarcity (43 %); lack of soil fertility (16.3 %), rabi season frost (10.9 %), and load-shedding (5.3 %). Intercropping could be a very useful practice that can enhance farm intensity above 100 %. However, it was very rare; only four localities reported it and with low quantities. Horticulture has been the main cultivation in the area's cropping pattern, with apple orchards the highest. Because of aridity and dwindling of aquifer potential, horticulture is gradually being replaced by field crops like vegetables, tobacco, and wheat, etc. On average, 62 % of the farmers have altered kharif-season cropping and 51.5 % have altered their rabi season crops.

Economic efficiency of tubewell farming has greatly suffered over time due to deterioration of the aquifer and its attendant problems of high irrigation costs. The initial capital cost of tubewell installation varies from area to area in the valley, depending upon factors such as DTW, total depth of bore, and quality of the material, etc. A total of 40.3 % of the tubewells ended at an initial installation cost of up to 50,000 rupees during 1981–90. This cost continued rising to between 50,000 and 1 lac rupees in the next 10 years, and culminated at around 8 lac rupees in the period 2005–08. Besides the hiking initial capital cost, the drying out of tubewells after short periods of service further exacerbates the economic crisis of tubewell irrigation. Driven largely by the flat rate policy in electricity tariff, only electric pump motors are being used, which, with increasing watertable drawdown, are also increasing in their horsepower specification. Initially in 1981–90, machines of 20–25 hp were sufficient for the purpose of water abstraction. Later, 30, 40, and ultimately 50 hp motors became the model types. Surely, a high-power motor costs more than a low-power one; for example, the average pump motor price during 1981–90 was around 25,000 rupees, which later increased to around 1 lac rupees during 2005–08. Likewise, maintenance and repair costs of the tubewells have also increased many-fold since 1981, when it was 1,000–2,000 rupees per month;

in the period 2005–08, it reached 8,000 rupees. All these factors add to the total cost of irrigation.

Though farm-income data may have limitations one way or the other, it is factual that the ever-escalating cost of irrigation has diminished farm profitability. In 1981–90, some 70 % of the farmers earned 5,000–20,000 rupees per acre per annum as gross income, with the rest of the farmers earning 20,000–60,000 rupees. Ten years later, during 1991–2000, the gross income in monetary terms showed an increase to a maximum range of 1 lac rupees. However, onwards to 2008, none surpassed this limit; rather, the percentage ratio of farmers earning a gross income of 1 lac rupees sharply declined to 3.4 % during 2001–04, and to 2.4 % during 2005–08. Since inflation rates have increased continuously, it means the gross income has also not generally improved. The net income situation is even more alarming. In 1981–90, its maximum limit was 30,000 rupees per acre per annum, with about half of the farmers earning up to only 5,000 rupees. During 1991–2000, the overall net income range is far wider; about 17 % of the farmers were earning up to 5,000, and only 1.4 % of them were earning 1 lac rupees per acre per annum. Until the year 2000, the net income, though small, was on the rise, but it gradually slumped in the following periods. This is why rural poverty is widespread in this society.

The study also aimed to uncover aquifer issues, including agricultural land use, cropping patterns, and farm income issues. It is an established fact that the cost of pumping use to grow proportionally with watertable depth. During 1981–90, watertable depth was an average of 89 ft in the valley, and dropped to 141 ft over the next 10 years at 5.2 ft per year. The watertable drawdown was phenomenal during 2001–04, with a maximum rate of 27 ft per year. This coincided with the prolonged and acute drought between 1998 and 2004. During 2005–08, the fall rate of the watertable was high, but it was significantly lower than the rate of 2001–04. Understandably, this relief in the watertable drawdown owed to the relatively high amount of rain and snow in 2005. Regarding the spatial diversity in watertable decline data among the constituting UCs/PCs, its great dependency was found upon the topography, location with respect to DADs and surface water drainage channels, and the farmers' memories and perceptions. Generally, we found that topography had a role in the DTW. In the high piedmont zones, the watertable was lower than the valley's low centre. Further, it was established that the per annum watertable decline rate was faster in areas of dense cultivation. Consequently, though due to population growth and technological advancement, new sections of the aquifer have been tapped and tubewell-irrigated agriculture on the whole has progressed, but simultaneously, the watertable has been shrinking since our analysis period began. This fact supports our hypothesis that tubewell-irrigated agriculture and the aquifer are affecting each other's sustainably towards negative end under the prevailing practices and strategies at all levels and ranks of society.

Excessive depletion of the watertable has initiated the phenomenon known as 'aquifer mining'. A few years ago, farmers used to drill not so deeply into the saturation zone because they were not expecting their tubewells to dry out so soon; and thus saved their money from useless investment. However, as they found the tubewells drying out sooner if they did not bore beyond the existing watertable, they now keep drilling further down unless and until any hard layer impedes drilling or the aquifer ends. This means the aquifer is being mined because, naturally, a reservoir hundreds of feet down cannot replenish within a reasonable time span, even if further abstraction is abandoned; i.e., that level is the permanent depletion of an aquifer. At such a depth, an aquifer would require centuries to recharge if the pores were not compressed too much. In the study area, that level has been reached as boring has gone down to more than 800 ft. Aside from the aquifer mining issue, the experts believe that when drilling reaches close to sea level, the phenomenon of 'seawater/or saline water intrusion' is highly likely, which would make the groundwater brackish. In fact, some chemical analyses have already indicated a slight brackishness in this area's groundwater.

The excessive aquifer drawdown has created a social equity problem. A great number of the resource-poor farmers have been prevented from the privilege of owning a tubewell. Their farms are either deserted or are being grown with purchased water, resulting in very nominal net income for them. During 1981–90, some 14.6 % of the irrigation farmers did not have their own tubewells. This ratio reduced to 6.7 % in 1991–2000, mainly motivated by the flat rate policy. However, in the following periods and through to the end of our data period, tubewell dropout cases have been increasing. Conversely, wealthy farmers have found good business. They install a number of tubewells, enjoy the flat rate policy, and sell the water at lucrative margins to those who do not have their own irrigation sources. Further, with the deepening of the watertable, the water yield of the tubewell has reduced because of the highly compressed and compact nature of the deeper permeable zone. This also serves as a compulsion for wealthy farmers to own many tubewells simultaneously. The electricity load-shedding is another reason for installation of many tubewells by a single owner.

The pH value of groundwater in the valley as a whole is within the normal range of 6.5–8.5 recommended by WHO as safe for human and plant use. Nevertheless, in some areas like Batezai, Alizai, and Yaro, the pH values are slightly above the normal range and it appears that water quality may become an issue in the near future. The concentration of TDS varies from 221 to 702 mg/L and is far below the recommended level of 1,000 mg/L. The perception of our participants was that water quality was relatively poor in the 1980s, but with deeper drilling it became good for some time after that. Since the year 2000, deterioration is again felt for domestic use of the water; for irrigation, it was never a limiting factor. However, salt color can be seen on top of the soil in some areas.

The survey has concluded that the role of Government has generally been small and ineffective in curtailing degradation of the groundwater resources for reasons like lack of political will, vested interests of the decision makers, ad hoc policies, and the poorly educated farming community, etc. Groundwater laws are either weak in themselves or are not implemented. In the area of on-farm water management, the Government has funded a number of farmers to improve their irrigation infrastructure, yet many are still deprived from any financial support to that end. The DADs built for aquifer recharge have not yielded promising results. The Government has no interference in crop selection, nor has it been found in extension services to farmers. The flat rate policy has been largely denounced by experts in the line department as being the cause of wasteful water use in irrigation works. Farmers and experts both have strongly recommended arrangement of area-suited efficient micro-irrigation technologies.

## 7.2 Key Findings

- Large household size, reflecting the joint-family system.
- Widespread illiteracy or substandard education among farmers.
- Non-farming commercial activities are common among the farming households, indicating per capita small farm sizes or low farm income.
- Based on the field survey data (durational data), the under-cultivation land (net-sown + current fallow) increased from 1981 until 2000, but later trends slowly but consistently downwards.
- Based on the satellite data (point data), the net-sown area increased from 1989 until 1996, decreased greatly in 2000, but increased again later in 2005 almost to the magnitude of 1996.
- Average irrigation farm-size per household increased steadily until 2004. Since then, it decreased slightly.
- Permanent/or temporary desertification of some acreage of land reported during 1991–95, which continues until the end of our study period but with a high magnitude until 2004 and relatively low magnitude during 2005–08.
- Farm fallowing has been practiced since before our study period (1981). The number of farm-fallowing farmers increased slightly during 1981–90 and abruptly during 1991–2000. No new farmer entered the list of farm-fallowing farmers during 2001–04, but a few entered the list during 2005–08.
- The main reason for farm fallowing is water scarcity, followed by lack of soil fertility, winter frost, and electricity load-shedding.
- The practice of intercropping is very rare.
- Due to widespread electrification and the flat rate policy in electricity tariff, all respondents owned only electric pumping machines, or were irrigating through such tubewells on a water purchase basis.
- All irrigation-related costs are increasing rapidly, primarily due to the rapid watertable depletion.
- Net farm income has not increased compatibly with general inflation and the cost of living in the country, hence rural poverty is encroaching.
- Grossly, the economic viability of irrigated agriculture is under severe threat.
- In many cases, changes in cropping patterns have emerged, either as a change from high-delta to low-delta crops or from low-value to high-value crops.
- The watertable is declining unabated everywhere in the valley, but at fluctuating annual rates.

- The watertable drawdown is creating a lull in cultivation ratio, increases in irrigation cost, and consequent decreases in farm income.
- Tubewell borings have reached to depths around 800 ft; thus, the phenomenon of aquifer mining has started.
- Water quality is, thus far, not affected significantly all across the valley.
- Official paperwork may exist; however, in practice, the Government is not actively interfering in tubewell installation or crop selection.
- Government support to farmers in the on-farm water management area is found in terms of funding of lining/piping the irrigation infrastructure (e.g. irrigation tanks and watercourses), but perhaps on a selected basis for influential farmers; many poor farmers are deprived in this regard.
- The DADs have not been able to make any major change in the aquifer discharge potential.
- The farmers are free in crop selection. The Government neither interferes in cropping pattern nor does it provide extension training to the farmers in this regard.
- The farmers are happy with the flat rate policy and wish it to continue; however, official experts consider it to be one of the main causes of water over-abstraction and consequent aquifer deterioration. Thus, they recommend calling it off.
- HEIS are considered a viable approach towards water conservation, but the role of Government in this regard would be imperative. It has been suggested that the Government provide such systems on subsidized rates to the farmers, in lieu of the subsidized electricity supply.

## 7.3 Suggestions

### *7.3.1 Structural and Technical Interventions*

- An integrated watershed rehabilitation and management pilot project should be initiated in the Pishin Lora basin. This way, the available surface water would be managed through appropriate techniques and structures such as pertinently designed small dams and ultimately made part of the aquifer instead of losing it through evaporation and surface runoff. The implementation of such a project would be helpful in checking groundwater depletion and the aquifer mining process.
- Micro-irrigation systems (drip, bubbler) apply water uniformly to crops, reduce seepage and evaporation losses, and thus can lead to high water productivity. These systems should be introduced for all tubewell-irrigated farms to effect complete conversion from flood irrigation to micro-irrigation.
- The available knowledge from the arid environments of the world should be accessed and utilized for the design of interventions for high-efficiency irrigation and precision farming technologies. Research should be directed to fulfilling

development needs. There is a strong need to reorient the focus of research institutions from traditional or discipline-oriented research to strategic and adaptive research.

- The Government should provide financial and technical support for HEIS (drip, bubbler, and sprinkler). If the drip irrigation systems are installed at 100 % cost to the Government, the whole of the electric tubewell-irrigated area can be converted to drip or sprinkler irrigation at a cost of around Rs. 17 billion (Ahmad 2006), which is around the total subsidy under the FRP for a period of 2 years. The private sector may also be encouraged to provide support in the micro-irrigation systems.
- Both water and energy resources are limited in Balochistan, therefore, any introduction of HEIS and precision farming should be aimed at reducing the power requirements. Thus, the interventions can be phased out as follows:
  - Introduction of energy-efficient pumping systems with overall system efficiency in the range of at least 70 % should be the entry point to improve energy use efficiency. This would include alternate sources of energy covering electric, diesel, and renewable energy.
  - The private sector may join hands whereby pump, power, and micro-irrigation producing companies may pool their resources to introduce energy-efficient pumping systems – micro-irrigation and precision farming. Introduction of HEIS must consider the farm as a whole, meeting the requirements of the farmers in terms of crop mix, soils, and climate. A start must be made from innovative and simple technologies, ultimately leading to standard and automated technologies in a gradual manner.

### 7.3.2 Policy and Institutional Reforms

- The highly subsidized electricity tariff for agriculture tubewells isolates the farmers' income from the costs of all inefficiencies at the farm level and prevents farmers from productive use of both water and energy. It also prevents the farmers from maintaining and looking after any sustainable intervention at the farm level. In addition to being a huge burden on the financial resources of the province, it is untargeted and non-functional to achieve any objective except providing relief to a few thousand farmers to subsidize their livelihood. Therefore, the subsidy must be called off. However, withdrawal of the subsidy is not going to be easily accepted by farmers and political icons given the lack of political will in provincial Government to address this issue on its merit. Therefore, it must be molded in a package acceptable to all stakeholders using a phase-wise approach agreed by all partners. The phase-wise recommendations are as follows:
  - In the first phase, the number of tubewells eligible for subsidy should be frozen and subsidy should be capped in nominal terms.

  - In the second phase, the power supply companies (QESCO) and the private sector should be encouraged to provide quality power supply to farmers so they are motivated to participate in the process of sharing the additional burden of electricity.
  - In the third phase, the subsidy should be gradually tapered off and farmers encouraged to adopt water conservation techniques such as drip irrigation, so that their energy requirement is reduced many-fold. The water conservation techniques would lead towards saving of the scarce groundwater resources of the province.

- Farmers are facing problems regarding a reliable supply of electricity. They complain of huge fluctuations in voltage, which is affecting the performance of electric motors and accessories. The prolonged periods of load-shedding and voltage fluctuations also affect the farmers' irrigation schedules, as they cannot meet their peak demands, which affects productivity of crops. QESCO must initiate schemes to improve infrastructure and management to ensure a reliable quality power supply to rural consumers. This would also help to create a discipline in payment of bills through ultimately metering the system.
- There is a need to create a 'groundwater authority' to regulate the aquifer and check its deficit.
- Pishin Lora is one of the overdrawn river basins of Balochistan. Lowering of the watertable and mining of groundwater is due to an increased number of tubewells and enhanced pumpage. Therefore, installation of more tubewells should be banned through regulatory control. Further, permits for installation of replacement tubewells should be linked to HEIS. A similar approach may be followed with the flat rate system of electricity tariff for tubewells, in that the power subsidy should be linked to the adoption of micro-irrigation systems.
- The policy for providing a license to install a tubewell in Balochistan should be linked to specifications for the depth of the tubewell and minimum efficiency levels for the installation of pumping systems. An increase of up to 50 % in energy use efficiency is possible by installing energy-efficient pumping systems, as pumping systems with overall pumping efficiency of over 75 % are now available in the market, in contrast to the current systems used by farmers that have efficiency of even less than 50 %.
- Care of Balochistan's water resources should be undertaken in a holistic and sustainable manner. In this regard, an integrated water resource management framework is an aspect that needs to be included while addressing the issue of water management, i.e. by improving cross-sectoral networking, and to improve understanding with different departments related directly and indirectly to the issue of water and other environmental sectors. The resource planning, management, and development should be responsive to basin boundaries as a hydrological unit. There should be incentives, regulatory controls, and public education, promoting economic efficiency, conservation of water resources, and protection of the environment.

- The success of any sustainable water resource management program is dependent on the availability of adequate and reliable data. It is a well understood principle all around the world that better water measurement equates to better water management. Adequacy and reliability of both surface and groundwater data is one of the major factors affecting planning, development, and management of water resources in the province. The water and land use data are hardly related to assessing reliability. Monitoring of floodwater and runoff is almost non-existent.

### 7.3.3 *Cropping Pattern Adjustment*

Water productivity is extremely low in the province, as it ranges between one-half to one-third of the potential water productivity of crops, fruits, and vegetables. Raising the water productivity in all the farming systems of Balochistan would have large impacts on water conservation, and improved productivity would lead towards increased profitability for irrigated agriculture. Switching from a more to a less water-consuming crop or switching to a crop with higher economic or physical productivity per unit of water consumed by evapotranspiration improves productivity of water. Thus, in tubewell-irrigated areas like Pishin Valley, the cultivation of onion should be discouraged and cultivation of crops like melons, water melons, cucumber, grapes, almonds, pistachio, pomegranate, fig, olive, pear, plum, etc. should be encouraged. Incentives should be given to the growers for growing water-productive crops.

### 7.3.4 *Social and Political Reforms*

Lack of awareness and motivation of farmers on the fact that water is a finite resource and is extremely scarce in Balochistan creates problems when trying to convince people of the importance of saving water for their own good, as well as for the good of the province and the nation. Therefore, enhancing the general literacy ratio through educational institutions, convening public debates, short awareness programs, and agricultural extension programs are essential for better soil management, irrigation scheduling, and crop selection, which would lead to more productivity of land and less consumption of water.

- There are major political challenges associated with the adoption of any new radical policy. Under traditional practices, members of the provincial and federal cabinets and local political bodies decide the projects to be started. They have their localized priorities and interests. The projects, rather than being based on technical, socio-economic, environmental, basin, or physiographic considerations – are designed to suit localized interests. The interests of the communities at large and ecosystem perspectives are therefore ignored. Although this trend seems difficult to amend in the prevailing tribal and democratic scenario, we have reasons to hereby appeal to our political leaders to become considerate to the recommendations of the experts in launching any developmental projects.

- There is a need to introduce subjects of water management and conservation at school, college, and university levels covering the resource picture of the province, including water, existing developments, and the innovative technologies available for efficient use of water and precision farming in the world.

## 7.4 Summary

This chapter summarized the findings of the study, which aimed to make a case for formulation of a revised strategy and effective institutional framework to enable the development of an economically viable local food production system. The chapter concluded that both tubewell farming and aquifer ecology are on a course for degradation, hence immediate policy shifts are required. With the new vision presented in this study, the agriculture and water crises can be minimized and slowed down; however, it may not be cured altogether. In my very sincere opinion, therefore, the concept of agriculture is inappropriate for most of the Balochistan uplands, including the study area. Therefore, instead of agriculture, which is, and most probably would continue to be, the highest water-consuming activity, the provincial economy should be switched over to some non-agricultural sectors. In saying that, I do not mean that agriculture should cease here altogether; however, the Government should divert its maximum incentives to sectors other than agriculture. For instance, there is great potential for minerals in the province, as well as for trade and industry, particularly with the launch of Gawadar Port. This port has promising potential for international trade, especially for the Central Asian States and China.

## Reference

Ahmad S (2006) Issues restricting capping of tubewell subsidy and strategy for introducing the smart subsidy in Balochistan, vol 2, no 1, TA-4560 (PAK). Supporting public resource management in Balochistan, Quetta

# Local Terms and Expressions Used

- **Bund**: literally in the Urdu language this means a dike, usually made of earth, to check a water stream. Contrary to a dam, a bund is a small structure made to divert a stream on to agricultural fields for irrigation purposes; to save river bank areas from flooding; or to store water for multiple uses, such as domestic, livestock, and irrigation.
- **Farm Size**: By farm size, we mean the area of holding irrespective of its location contiguously or fragmentally.
- **Karez**: An underground water conveyance system from one or several mother wells dug at the foothills to lowlands over a distance of several hundred meters. Karezes are the typical means of irrigation in the dry mountainous zones of Balochistan province of Pakistan. Being underground, the karezes help reduce water loss through evaporation.
- **Kharif Season**: Summer to autumn season – May to October in Pakistan.
- **Kharif Crops**: Crops grown in kharif season – such as maize, rice, melons.
- **Khushkaba Farming**: Dry farming, i.e. farming under direct precipitation only.
- **Load-shedding**: Breakdown of electric supply.
- **Lora**: A rainwater stream flowing downslope from highlands. The word is derived from a Pashto language word 'Lwarha', which means 'height'.
- **Nala**: A small rainwater stream; a sewerage drain; or any other small water channel.
- **Patwar Circle (PC)**: The total land revenue area, comprising several villages, under one patwari, who is a revenue official at field level in charge of maintaining land records, conducting property transactions, surveying crops, and collecting revenue.
- **Rabi Crops**: Crops grown during rabi season – such as wheat, barley, cumin.
- **Rabi Season**: Winter to spring season – November to April in Pakistan.
- **Sailaba Farming**: Urdu translation for the word 'flood' is 'sailab'. Sailaba farming is similar to spate farming (or spate irrigation) with a minor difference, and that is, in the latter, flood water is diverted to crops by large dikes across hill torrents;

A.S. Khattak, *Mutual Sustainability of Tubewell Farming and Aquifers: Perspectives from Balochistan, Pakistan*, Advances in Asian Human-Environmental Research, DOI 10.1007/978-3-319-02804-0, 

in the former, small guide structures are constructed on hill slopes to divert rain-water from other fallow areas to cropped fields.

- **Shamilat**: Common lands of a village or tribe.
- **Tehsil**: The second-order administrative subdivision of a district.
- **Union Council (UC)**: The lowest political unit comprising a few (four to five) villages. In the Devolution of Power Plan, 2001, a UC was supposed to be politically represented by an elected body of several members headed by a UC nazim (nazim means 'administrator' in local language).
- **Village**: A demarcated rural landscape of a population less than 5,000 for which separate revenue records, including a cadastral map, is maintained.

# Appendices

## Appendix A: Mosaic of the Toposheets (R.F. 1:150,000,000) Covering Pishin Valley

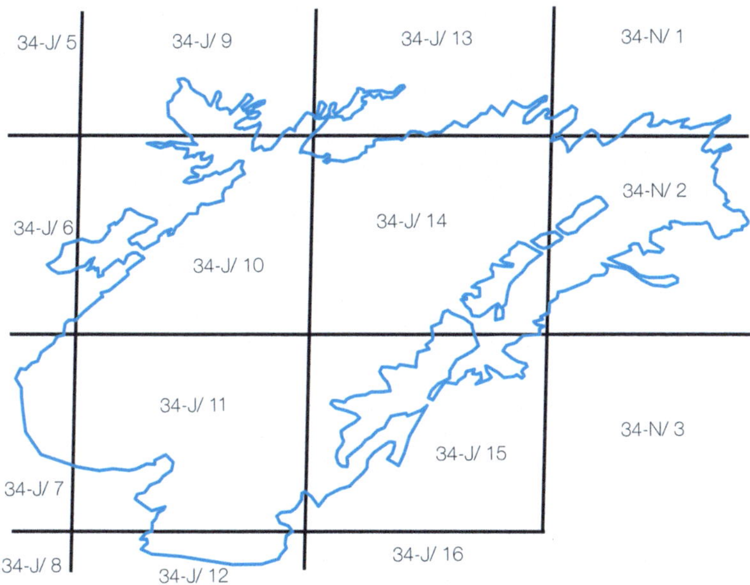


A.S. Khattak, *Mutual Sustainability of Tubewell Farming and Aquifers: Perspectives from Balochistan, Pakistan*, Advances in Asian Human-Environmental Research,
DOI 10.1007/978-3-319-02804-0, 

## Appendix B: Lithology of Pishin Valley

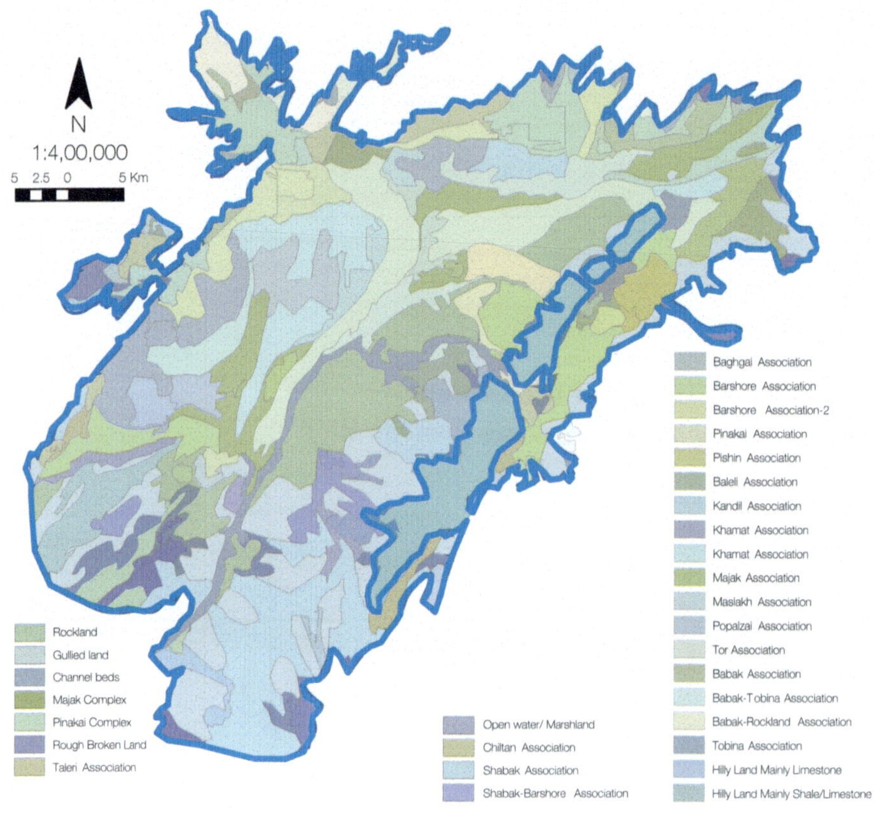


Source: GSP

# Appendix C: Landcover Statistics of Pishin Valley Calculated from Landsat Images

| | | Agriculture | | | | |
|---|---|---|---|---|---|---|
| UC/PC | Year | Total acreage | % of total area of the UC/PC | Bare soil/ Settlement/Hard rock (acre) | Barren land with salt color (acre) | Wetlands (acre) |
| Ajram Shadezai | 1989 | 88.51 | 0.05 | 119,841.78 | 66,788.74 | 77.1 |
| 186,576 acres | 1991 | 93.53 | 0.05 | 112,253.82 | 74,143.3 | 340.6 |
| (755 km$^2$) | 1996 | 109.1 | 0.06 | 100,894.89 | 85,373.6 | 303.62 |
| | 2000 | 0.4 | 0 | 158,658.71 | 27,683.85 | 92.32 |
| | 2005 | 98.6 | 0.05 | 123,102.1 | 63,155.55 | 219.4 |
| Alizai | 1989 | 330.35 | 3.84 | 4,734.1 | 3,058.4 | 511.21 |
| 8,596 acres | 1991 | 741.28 | 8.63 | 6,677.66 | 539.3 | 692.63 |
| (4.8 km$^2$) | 1996 | 803.59 | 9.35 | 4,197.34 | 3,014.1 | 555.7 |
| | 2000 | 516.86 | 6.02 | 6,588.28 | 655 | 796 |
| | 2005 | 789.66 | 9.19 | 6,258.23 | 1,002.56 | 545.9 |
| D. Khanzai | 1989 | 1,489.25 | 18.4 | 2,003.46 | 4,146.2 | 531.67 |
| 8,120 acres | 1991 | 3,602.62 | 44.37 | 3,791.11 | 176.9 | 570.67 |
| (32.9 km$^2$) | 1996 | 4,712.94 | 58.05 | 1,356.78 | 1,489.6 | 614.92 |
| | 2000 | 2,542.16 | 31.11 | 4,611.18 | 298.8 | 662.1 |
| | 2005 | 3,317.44 | 40.86 | 3,222.6 | 1211.8 | 367.8 |
| Batezai | 1989 | 1,782.28 | 10.76 | 12,150.82 | 2,403.25 | 307.7 |
| 16,571 acres | 1991 | 2,725.66 | 16.45 | 13,223.24 | 137.8 | 523.9 |
| (67.1 km$^2$) | 1996 | 3,383.59 | 20.42 | 12,586.9 | 578.74 | 24.11 |
| | 2000 | 2,074.4 | 12.52 | 14,414.7 | 26.5 | 6 |
| | 2005 | 3,636.5 | 21.95 | 11,536.36 | 1,136.14 | 262.18 |
| Gangalzai | 1989 | 262.35 | 5.04 | 2,166.2 | 2,645 | 5.21 |
| 5,210 acres | 1991 | 652.73 | 12.53 | 4,154.79 | 407.52 | 79.4 |
| (21.1 km$^2$) | 1996 | 809.42 | 15.54 | 1,547.54 | 2,796.73 | 0 |
| | 2000 | 665.49 | 12.77 | 3,843.1 | 683.21 | 0.35 |
| | 2005 | 1,006.27 | 19.32 | 2,658.89 | 1,542 | 3.1 |
| Gulistan | 1989 | 8,305.25 | 14.41 | 1,1649.63 | 3,0540.8 | 356.7 |
| 57,654 acres | 1991 | 11,979.1 | 20.78 | 27,346.26 | 14,766.65 | 783.9 |
| (233.32 km$^2$) | 1996 | 14,800.89 | 25.67 | 9,607.32 | 29,896.63 | 1,416.5 |
| | 2000 | 11,401.34 | 19.78 | 30,255.64 | 12,751.6 | 1,227.13 |
| | 2005 | 15,924.92 | 27.62 | 22,024 | 18,541.1 | 1,164.04 |
| Huramzai | 1989 | 1,195.8 | 10.82 | 3,559.62 | 6,457.8 | 28.5 |
| 11,056 acres | 1991 | 1,786.78 | 16.16 | 7,485.57 | 1,744.8 | 88.74 |
| (44.7 km$^2$) | 1996 | 2,251.62 | 20.37 | 2,022 | 6,729.6 | 68.4 |
| | 2000 | 1,877.21 | 16.98 | 4,125.9 | 5,398.22 | 109.64 |
| | 2005 | 2,531.12 | 22.89 | 5,605.41 | 2,877.11 | 42.43 |
| Karbala | 1989 | 161.78 | 0.63 | 2,945.18 | 22,859.13 | 52.8 |
| 25,953 acres | 1991 | 347.45 | 1.34 | 4,417.85 | 2,1193.2 | 51.74 |
| (105 km$^2$) | 1996 | 305.89 | 1.18 | 1,314.18 | 24,111.5 | 303.8 |
| | 2000 | 103.81 | 0.4 | 3,622.46 | 22,198.2 | 15.64 |
| | 2005 | 421.33 | 1.62 | 6,735.7 | 18,761.4 | 34.2 |

(continued)

| UC/PC | Year | Agriculture: Total acreage | Agriculture: % of total area of the UC/PC | Bare soil/ Settlement/Hard rock (acre) | Barren land with salt color (acre) | Wetlands (acre) |
|---|---|---|---|---|---|---|
| Maizai 15,831 acres (64.1 km²) | 1989 | 2,188.91 | 13.83 | 1,880.83 | 11,773.3 | 38.33 |
| | 1991 | 2,869.67 | 18.13 | 869.69 | 11,975 | 159.4 |
| | 1996 | 3,087.66 | 19.51 | 522.41 | 11,992.6 | 29.5 |
| | 2000 | 3,040.47 | 19.21 | 1,863.2 | 10,742.3 | 174.2 |
| | 2005 | 3,254.38 | 20.56 | 3,613.37 | 8,875.4 | 88 |
| Malakyar 12,685 acres (51.3 km²) | 1989 | 1,809.98 | 14.27 | 5,852.42 | 3,245.83 | 1,847.14 |
| | 1991 | 3,457.77 | 27.26 | 6,895.6 | 362.62 | 2,004.4 |
| | 1996 | 4,277.8 | 33.73 | 4,412.24 | 2,367 | 1,654 |
| | 2000 | 3,773.86 | 29.75 | 5,621.78 | 1,642.2 | 1650.15 |
| | 2005 | 4,495.02 | 53.44 | 6,050.89 | 1,048.9 | 1,089.8 |
| Manzaki 12,098 acres (49 km²) | 1989 | 243.86 | 2.02 | 9,737.4 | 17,733.9 | 570.42 |
| | 1991 | 724.84 | 6 | 1,0546.85 | 261.9 | 654 |
| | 1996 | 773.67 | 6.4 | 8,212.25 | 2,465.9 | 623.87 |
| | 2000 | 605.89 | 5 | 9,221.72 | 1,632.2 | 577.81 |
| | 2005 | 954.99 | 7.9 | 9,395.39 | 1,300.7 | 447.26 |
| Manzari 14,262 acres (57.72 km²) | 1989 | 954.97 | 6.7 | 5,298.45 | 7,484.4 | 602.12 |
| | 1991 | 2,486.86 | 17.44 | 9,676.53 | 1,225.5 | 928.72 |
| | 1996 | 3,063.13 | 21.48 | 5,084.35 | 5,432.9 | 706 |
| | 2000 | 1,643.93 | 11.53 | 10,265.47 | 1,318.9 | 1,009.7 |
| | 2005 | 2,865.12 | 20.1 | 8,678.47 | 2,150.1 | 568.24 |
| Pishin City 8,109 acres (32.82 km²) | 1989 | 413.45 | 5.1 | 6,206.46 | 1,508.3 | 1.2 |
| | 1991 | 639.45 | 7.9 | 7,212.92 | 283.5 | 13.52 |
| | 1996 | 852.03 | 10.51 | 5,388.12 | 1,864.3 | 7.3 |
| | 2000 | 541.72 | 6.7 | 7072.42 | 467.3 | 0 |
| | 2005 | 1,406.59 | 0.17 | 5,874.78 | 817.5 | 10.6 |
| N. Malizai 27,268 acres (110.35 km²) | 1989 | 1,258.63 | 4.62 | 1,1487.31 | 1,4558.9 | 45.76 |
| | 1991 | 2,805.87 | 10.3 | 22,055.49 | 2,467.5 | 38.23 |
| | 1996 | 4,042.33 | 14.82 | 8,186.26 | 15,070.1 | 27.23 |
| | 2000 | 1,270.79 | 4.66 | 19022.77 | 6,901.73 | 41.74 |
| | 2005 | 4,702.3 | 17.25 | 10,819.21 | 11,731 | 15.3 |
| Q. Abdullah 33,165 acres (134.2 km²) | 1989 | 3,457.57 | 10.43 | 19,867 | 7,190.2 | 2,817.4 |
| | 1991 | 5,915.35 | 17.84 | 22,347.05 | 1,209.4 | 3,883.24 |
| | 1996 | 6,218.74 | 18.75 | 16,082.63 | 6,782.13 | 3,882.4 |
| | 2000 | 3,962.52 | 11.95 | 2,4735.8 | 487.2 | 3,714.8 |
| | 2005 | 4,358.2 | 13.14 | 20,737.4 | 5,825.14 | 2,244.04 |
| Saranan 23,408 acres (94.73 km²) | 1989 | 36.72 | 0.16 | 11,607.73 | 11,822.7 | 9.64 |
| | 1991 | 80.21 | 0.34 | 21,205.92 | 2178.3 | 32.62 |
| | 1996 | 192.64 | 0.82 | 9,037.48 | 14,220 | 0 |
| | 2000 | 18.71 | 0.08 | 20,932.12 | 2,423.4 | 0 |
| | 2005 | 255.1 | 1.09 | 14,337.67 | 8,811.1 | 3.9 |
| Segi 90,958 acres (368.1 km²) | 1989 | 857.24 | 0.94 | 46,846.95 | 24,463.8 | 163.39 |
| | 1991 | 2,732.1 | 3 | 72,216.84 | 11,615.2 | 173.32 |
| | 1996 | 3,444.48 | 3.8 | 56,662.1 | 28,851.55 | 65.68 |
| | 2000 | 2,228.97 | 2.45 | 75,488.43 | 11,459 | 2.47 |
| | 2005 | 6,529.62 | 7.18 | 57,731 | 26,271.63 | 425.84 |

(continued)

| UC/PC | Year | Agriculture: Total acreage | Agriculture: % of total area of the UC/PC | Bare soil/ Settlement/Hard rock (acre) | Barren land with salt color (acre) | Wetlands (acre) |
|---|---|---|---|---|---|---|
| Simzai | 1989 | 769.32 | 5.8 | 5,961.2 | 6,296.8 | 353.46 |
| 13,330 acres | 1991 | 1,502.89 | 11.3 | 10,128.7 | 1,212.2 | 541.36 |
| (54 km$^2$) | 1996 | 1,770.97 | 13.29 | 5,534.22 | 5,635.44 | 406.69 |
| | 2000 | 1,614.75 | 12.1 | 9,852.59 | 1,386.2 | 457.32 |
| | 2005 | 1,996.96 | 15 | 8,599.1 | 2,381.9 | 351.78 |
| Tora Shah | 1989 | 2,273 | 16.2 | 9,848.89 | 1,809.6 | 139.49 |
| 14,036 acres | 1991 | 3,721.47 | 26.5 | 9,944.39 | 61.31 | 389.24 |
| (56.8 km$^2$) | 1996 | 4,292.55 | 30.58 | 9,042.7 | 688.5 | 50.9 |
| | 2000 | 2,809.26 | 20 | 1,0367.7 | 578.1 | 246.22 |
| | 2005 | 3,736.84 | 26.6 | 8,882.21 | 1,024.7 | 392.13 |
| Yaro 12,194 acres | 1989 | 89.52 | 0.73 | 12,148.38 | 2,294.7 | 7.41 |
| (49.35 km$^2$) | 1991 | 157.3 | 1.3 | 13,055.31 | 1,255.8 | 50.7 |
| | 1996 | 172.48 | 1.4 | 10,547.47 | 3,313.8 | 342.5 |
| | 2000 | 69.44 | 0.57 | 13,247.8 | 1,008 | 0.3 |
| | 2005 | 350.34 | 2.9 | 8,575.51 | 3,248.7 | 19.2 |
| **Pishin Valley total area** | | **5,97,079.16 acres (2,416 km$^2$)** | | | | |

Source: Saeed, 2012 (Derived statistics from Landsat data)

# Appendix D: Groundwater Basins of Balochistan Province

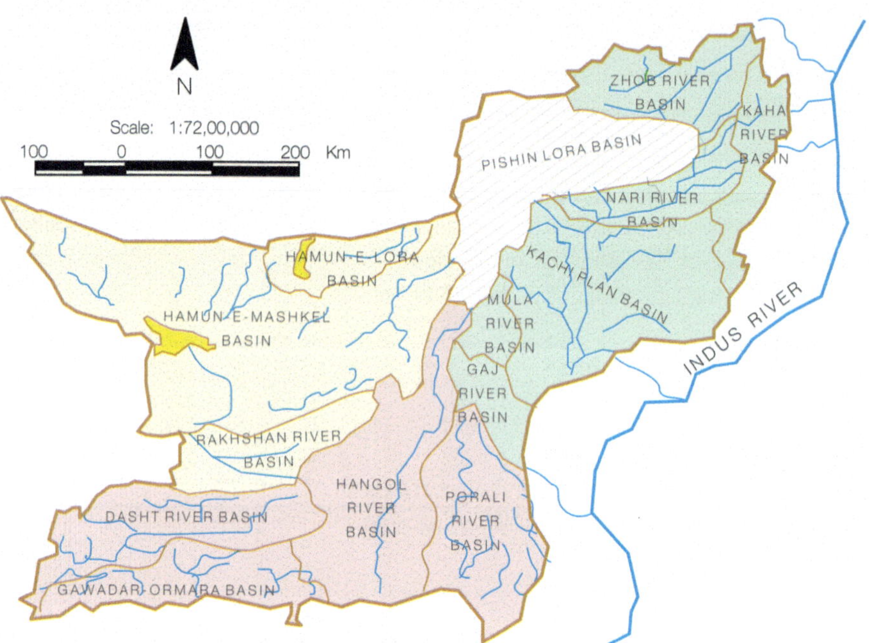


Source: Directorate General of Hydrology, WAPDA

# Appendix E: Photograph of a Deserted Fruit Orchard in Pishin Valley

Source: Saeed (2012)

# Appendix F: Questionnaire A: Farmers

Topic: - Inter-Effects of Tubewell Irrigated Agriculture and the Aquifer Potential in Balochistan: Case Study of Pishin Valley, 1981-2008.

File No. _____Date ________Interviewer ________________Respondent (name, age) _______________

Village _________________UC/PC ________________ Tehsil _____________District ______________

**1. Household Information**

Family size_____ No. of fulltime farmers_____ Farmers' education ______ Nonfarm Occupations ______

Farm share in income: 2008 ___________2004___________2001 ___________1991___________1981

**2. Land use (acres)**

Total holding: 2008______________2004_____________2001____________1991___________1981

Under irrigated cultivation: 2008___________ 2004___________2001___________1991__________1981

Under dy cultivation: 2008____________20054___________2001___________1991___________1981

Cultivable-waste land: 2008____________2004____________2001__________1991______ ____1981

Cultivated land deserted _______Since the year _______Why: 1 _______________2_______________

Use of the deserted land ____________Average year fallow (% of under cultivation): Kharif season______

Rabi season______ from when fallowing practiced_____ Why fallowing: 1___________2___________

Avg. year % land intercropped: Kharif ____Rabi ____Crops involved: Kharif __________Rabi _________

**3. Water Availability**

Tubewells owned: 2008___ ____________2004____________2001____________1991____________ 1981

Tubewells dried: 2008 _______________2004____________2001____________1991___ __________1981

Bore depth (ft): 2008______________2004_____________2001____________1991_____________1981

Watertable depth: 2008______________2004____________2001____________1991____________1981

Did the nearest DAD raised watertable: Yes / No; Other uses of DAD ________________________

**4. Water Quality**

For humans –good/fair/poor: 2008 _________2004 _________2001 _________1991 _________1981

For laundry –good/fair/poor: 2008 _________2004 _________2001 _________1991 _____ ____1981

For irrigation –good /fair/ poor: 2008 ________2004 _________2001 _________1991___ ___ ___1981

Salt colour on surface: Yes / No; Effects on crops ____________________________________

*5. Irrigation cost*

Bore + casing cost: 2008 __________2004 __________2001 ___________1991 ___________1981

Motor horsepower: 2008 __________2004 __________2001 ___________1991 ___________1981

Motor price: 2008 __________2004 _____ _____2001 ___________1991 ___________1981

Monthly O & M cost: 2008 _____ ___2004 ___________2001 ___________1991 ___________1981

**6. Irrigation Technology**

No. of alive tubewells ______ subsidized _____ Distance of nearest tubewell ___Does govt check new boring: Yes / No; How many your tubewells are licensed ___ permitted motor power ____ Daily operation (hrs): Kharif___Rabi ____ Do you sell water to non-owner farmers: Yes / No; Do you use HEIS: Yes / No; If yes, type: Drip / Bubbler; Installation year ____ Who funded________ HEIS effect on production: increased / decreased / no change; Crops & area served by HEIS_____ If HEIS not used, Why: _________ Water tank lined: Yes / No; % of lined watercourses ___ % of piped watercourses ____ Who funded ____

**7. Farm-Income** (annual average per acre)

Gross income: 2008 __________2004 _____________2001 ___________1991 ___________1981

Net income: 2008 ___________2004 _____________2001 ___________1991 ___________1981

**8. Cropping Pattern**

Principle crops (% of area): Kharif _________________________ Rabi _______________________

Water scarcity driven cropping change: Kharif ________________ Rabi _______________________

Does govt. check crop selection: Yes/ No; If yes, which crops restricted_________________________

Does govt. give incentives for certain crops: Yes/ No, If yes, which crops ________Which incentives_____

Have you been invited to training on crops selection: Yes / No; If yes, how many you attended ________

Name the training agencies ___________________________________________________________

**9. Views About the 'Flat Rate Policy' in Electricity Tariff**

1. Is the Flat Rates Policy causing water overuse: Yes/ No

2. Should Govt. drop the Flat Rate Policy: Yes/ No

3. How would you react if the Policy is dropped: i) Not mind ☐ ii) Protest unconditional ☐

Demand alternatives ☐, like ________________________________________________________

4. What you may do if had to pay full electricity bill: i) Continue with electric motor ☐

ii) Change to diesel machine ☐ iii) Change to rain farming ☐ iv) Abandon farming ☐

5. Is the current Flat Rate of electricity is: High / Fair / Low; Justify answer__________________

**10. Respondent's Suggestions for Water Conservation:** ___________________________________

**Interviewer's Signature** ___________________**Respondent's Signature** ______________________

## Abbreviations and Local terms

**DAD**: Delay Action Dam
**HEIS**: High Efficiency Irrigation Systems
**Kharif crops**: Summer season crops
**Rabi crops**: Winter season crops

# Appendix G: Questionnaire B: Key Informants

Topic: Inter-Effects of Tubewell Irrigated Agriculture and the Aquifer Potential in Balochistan: Case Study of Pishin Valley, 1981-2008.

File No. ____Date _______Informant's name ____________Designation ____________________________

Department____________________________________________________________________________

**Instructions:** Please tick one box affront each answer.

[0] Shows: not agreed / not significant/ or untrue

[1] Shows: highly agreed / highly significant/ highly true

[2] Shows: moderate or fair agreement/ significance/ truth

[3] Shows: slight agreement/ significance/ truth

**Note:** If applicable, a same level of agreement/ significance may be marked for several answers.

**A- Agricultural Development**

1. Why are there vast extents of cultivable land barren in Pishin Valley?
   i. Lack of irrigation water: .............. [0] [1] [2] [3]
   ii. Abundance of land disputes:.......... [0] [1] [2] [3]
   iii. Lack of population/ laborers:.......... [0] [1] [2] [3]
   iv. Lack of interest of owners: ........... [0] [1] [2] [3]
   v. Poverty of owners:...................... [0] [1] [2] [3]
   vi. Abundance of state lands: ............ [0] [1] [2] [3]
   vii. Any other reason: ____________________________________________

**2. What do you suggest for sustainable agriculture (social, economic, ecologic) in the area?**
   i. Control groundwater abstraction by tubewell licensing: [0] [1] [2] [3]
   ii. Subsidized provision of HEIS: ........................... [0] [1] [2] [3]
   iii. Emphasis on intercropping: ................................ [0] [1] [2] [3]
   iv. Adoption of low delta crops:............................... [0] [1] [2] [3]
   v. Protection/development of recharge zones in watershed: [0] [1] [2] [3]
   vi. Emphasis on traditional Sailaba & Khushkaba farming: [0] [1] [2] [3]
   vii. Any other suggestion: ____________________________________________

**3. Which changes water shortage has brought in cropping pattern?**
   i. Shift from orchards to field-crops: ......................... [0] [1] [2] [3]
   ii. Shift from low value crops to high value crops:.......... [0] [1] [2] [3]
   iii. Shift from high delta crops to low delta crops: ........... [0] [1] [2] [3]
   iv. Any other change: ____________________________________________

**4. Which steps govt. taking to popularize low delta crops?**

i. Restricts cultivation of high delta crops:............... [0] [1] [2] [3]
ii. Subsidizes seeds, etc. of low delta crops:.............. [0] [1] [2] [3]
iii. Trains and educate farmers for low delta crops:...... [0] [1] [2] [3]
iv. Increases support price for low delta crops:......... [0] [1] [2] [3]
v. Provides special marketing for low delta crops:.... [0] [1] [2] [3]
vi. Any other step: ______

**B- Aquifer Issues**

5. What are reasons for the quicker watertable decline in Pishin Valley?

i. Precipitation reduction due to climate change: .......... [0] [1] [2] [3]
ii. Elimination of recharge zones by settlements expansion: [0] [1] [2] [3]
iii. Enormous development of tubewell farming: ............. [0] [1] [2] [3]
iv. Misuse of water in irrigation system: .............. [0] [1] [2] [3]
v. Any other reason: ______

6. What do you suggest to avert quick fall of watertable?

i. Change from agriculture to industry or else less water intensive landuse: [0] [1] [2] [3]
ii. Construction of more DADs & other recharge structures: ................ [0] [1] [2] [3]
iii. Use of HEIS and better irrigation scheduling: .............................. [0] [1] [2] [3]
iv. Complete ban on further boring for irrigation use........................... [0] [1] [2] [3]
v. Any other suggestion: ______

**C- Water Management Strategies**

7. What impedes enforcement of groundwater regulations?

i. Lack of political will ............................... [0] [1] [2] [3]
ii. Poor accessibility of the area........................ [0] [1] [2] [3]
iii. Weakness of enforcing agencies..................... [0] [1] [2] [3]
iv. Corruption on the part of enforcing agencies...... [0] [1] [2] [3]
v. Inadequacy of the regulations:........................ [0] [1] [2] [3]
vi. Any other reason: ______

8. Have Delay Action Dames proved useful as deemed? Yes [ ] No [ ] . ----If yes, how?

i. Much effective in long duration aquifer recharge: [0] [1] [2] [3] [5]
ii. Much effective in short term aquifer recharge:... [0] [1] [2] [3] [5]
iii. Marginally effective in aquifer recharge: ......... [0] [1] [2] [3] [5]
iv. Effective in flood control:........................... [0] [1] [2] [3] [5]
v. Useful as non irrigational water uses: ........... [0] [1] [2] [3] [5]
vi. Any other benefits: ______

9. It was how where the DADs failed to come up to expectations?

i. Quick sedimentation: .................. [0] [1] [2]
ii. Choking of subsurface pores quickly: [0] [1] [2]
iii. Any other shortfall: ______

10. What constrains adoption of High Efficiency Irrigation Systems?

i. High operation & maintenance cost: ........................ [0] [1] [2] [3]

ii. Lower production than flood irrigation:...................... [0] [1] [2] [3]

iii. Availability issues of the system & spare parts: .......... [0] [1] [2] [3]

iv. Lack of consciousness for water conservation: ............ [0] [1] [2] [3]

v. Lack of skill in using HEIS: .................................. [0] [1] [2] [3]

vi. Cheap availability of water for flood irrigation due to FRP: [0] [1] [2] [3]

vii. Any other constraint: ____________________

11. What do you suggest to make High Efficiency Irrigation Systems popular?

i. Subsidized supply of HEIS: ........................... [0] [1] [2] [3]

ii. Withdrawal of the Flat Rates Policy: .............. [0] [1] [2] [3]

iii. Making use of HEIS a condition for power subsidy: [0] [1] [2] [3]

iv. Any other suggestion: ____________________

12. Do you think the F R P. is the cause of water misuse? Yes [ ] No [ ]

13. Do you support dropping FRP? Yes [ ] No [ ]. If no, why__________

14. Which motives drive the Flat Rates Policy?

i. Vested interests of decision makers:........................ [0] [1] [2] [3]

ii. Fear of public resentment against dropping FRP:......... [0] [1] [2] [3]

iii. WAPDA's inabilities in thefts checking & bill collection: [0] [1] [2] [3]

iv. Farmers' welfare; to reduce their irrigation cost: ......... [0] [1] [2] [3]

v. Any other motive: ____________________

15. Do you think dropping FRP may cause public resentment? Yes [ ] No [ ] - --If yes, how?_______

16. Suggest socially / politically acceptable alternatives/ modifications in the FRP: __________

17. Informant's open note on the subject issues: ____________________

**Interviewer's signature:** ____________ **Informant's signature:** ____________

# Literature Studied (un-cited)

Abdullah M, Farooq S, Qureshi JD (1996) Declining groundwater resources of Balochistan – measures of replenishment. Proceedings of the regional workshop on artificial groundwater recharge, 10–14 June 1996. Pakistan Council of Research in Water Resources, Quetta

Ahmad S (2007) Why diagnostic process is essential for designing research for development – improving performance of irrigated agriculture in Pakistan, vol 1, no 3, TA-4560 (PAK), Quetta

Arif SM (1991) Agricultural economy of Balochistan: selected papers. Diffusion of tube well technology and Karez Abandonment in Balochistan: a socio-economic perspective. Quetta Printing Press, Quetta, pp 25–38

Baber R (1998) A comparative statistical analysis of agricultural development of the Quetta and Pishin Areas, Balochistan, Pakistan. Published in 'Studies in Pakistan Geography'. Israr-ud-Din (ed) Department of Geography, University of Peshawar, Peshawar

Barrow C (1987) Water resources and agricultural development in the tropics. Longman, Harlow

Bhutta MN, Ramazan M, Hafeez CA (2002) Groundwater quality and availability in Pakistan. In: Proceedings of the seminar on "Strategies to address the present and future water quality issues", 6–7 Mar 2002. Pakistan Council of Research in Water Resources, Islamabad, Pakistan

BMIADP (1990) Artificial groundwater recharge: final Report. Balochistan minor irrigation and agriculture development Project, Halcrow -ULG/Euroconsult

Bradley WP, Mark WL (1989) Irrigation development: social science perspective. Readings and notes of workshop held 25 Feb–02 Mar 1989, Lahore

Chaudhary MA (1989) Agricultural development and public policies: with special reference to Balochistan. Izharsons, Lahore

Ewert FA, van Ittersum MK, Bezlepkina I et al (2005) Development of a conceptual framework for integrated analysis and assessment of agricultural systems in SEAMLESS-IF, SEAMLESS Report no. 1, SEAMLESS integrated project. Accessed at: http://ageconsearch.umn.edu/bitstream/9287/1/re050001.pdf. Accessed 28 July 2009

Hassan Q, Garg NK (2003) In: Rema D, Naved A (eds) Management of water resources: an integrated perspective. Printed in Water and Wastewater Perspectives of Developing Countries, conference proceedings. Anamaya Publishers, New Delhi

Hecht R (1991) Land and water rights and the design of small-scale irrigation projects: the case of Balochistan. Water Resources Journal, June, 1991, Economic and social commission for Asia and the Pacific, UNO

Kahlown MA, Ashraf M (2002) Water management strategies under drought conditions. Published in Proceedings of the SAARC workshop on drought and water management strategies, 16–18 Sept 2002, Lahore, Pakistan, pp 47–57

A.S. Khattak, *Mutual Sustainability of Tubewell Farming and Aquifers: Perspectives from Balochistan, Pakistan*, Advances in Asian Human-Environmental Research, DOI 10.1007/978-3-319-02804-0, 

Majeed A, Ali S (2006) Water management practices in Balochistan (unpublished), IUCN, Balochistan Programme

Mian BA, Abdullah M (2000) Impacts of groundwater development by Tubewell Technology on Karez in Balochistan, Pakistan. Published in Proceedings of the first international symposium on Qanat, pp 127–136, 8–11 May 2000, Yazd

Morton J, van Hoeflanken H (1995) Some findings from a survey of flood irrigation schemes in Balochistan, Pakistan. J Water Resour. Published by the Economic and Social Commission for Asia and the Pacific, UNO

Nazir A (1995) Groundwater resources of Pakistan. Shehzad Nazir Publisher, Lahore

Oosterbaan JR (1983) Modern interferences in traditional water resources in Balochistan. Water Resour J Economic and Social Commission for Asia and the Pacific, UNO

Perkins M, Birch DR (1999) Groundwater recharge in the Quetta Valley and surrounding areas: prospective techniques and their potential for mitigation of the decline in watertable levels. Proceedings of the Regional Workshop on Artificial Groundwater Recharge, 10–14 June 1996, Quetta, pp 14–20

Rosenzweig C, Hillel D (2005) Climate change, agriculture and sustainability. In: Lal R, Uphoff N, Stewart BA, Hansen DO (eds) Climate change and global food security. Taylor & Francis, Boca Raton, pp 243–268

Scanlon BR, Keese KE, Flint AL, Flint LE, Gaye CB, Edmunds WM, Simmers I (2006) Global synthesis of groundwater recharge in semiarid and arid regions. Wiley, New York, Accessed at: http://www.beg.utexas.edu/staffinfo/Scanlon_pdf/ScanlonHJ02.pdf. Accessed 22 July 2013

Schoute JFT, Finke PA et al (1995) Scenario studies for the rural environment. Kluwer Academic Publishers, Dordrecht

Schwartz WF, Zhang H (2003) Fundamentals of groundwater. Wiley, New York

Todd DK (1980) Groundwater storage and artificial recharge. UN department of economics and social affairs, natural resources/Water Series, no. 2

World Bank (2005) Shaping the future of water for agriculture: a sourcebook for investment in agricultural water management, Washington, DC. Available at: http://siteresources.worldbank.org/INTARD/Resources/Shaping_the_Future_of_Water_for_Agriculture.pdf. Accessed 10 Sep 2013

# Index



A.S. Khattak, *Mutual Sustainability of Tubewell Farming and Aquifers: Perspectives from Balochistan, Pakistan*, Advances in Asian Human-Environmental Research,
DOI 10.1007/978-3-319-02804-0, 

**R**



**S**

The manufacturer's authorised representative in the EU is Springer Nature Customer Service Centre GmbH, Europaplatz 3, 69115 Heidelberg, Germany. If you have any concerns regarding our products, please contact ProductSafety@springernature.com

Printed and bound by CPI Group (UK) Ltd, Croydon, CR0 4YY

15/07/2026

02167649-0002